THE PHYSIOLOGY OF
THE LOCUST EAR (I-III)

II

Denne afhandling
er af det naturvidenskabelige fakultet sråd
ved Københavns universitet antaget til offentlig at forsvares
for den filosofiske doktorgrad
København, den 29. marts 1971
Morten Lange
dekan

ISBN 978-3-662-39252-2   ISBN 978-3-662-40271-9 (eBook)
DOI 10.1007/978-3-662-40271-9

Forsvaret finder sted tirsdag den 15. juni 1971 kl. 14 præcis
i Annexauditorium A, Studiestræde 6, o.g.

The Physiology of the Locust Ear

I. Frequency Sensitivity of Single Cells in the Isolated Ear
II. Frequency Discrimination Based upon Resonances in the Tympanum
III. Acoustical Properties of the Intact Ear

AXEL MICHELSEN

Reprint from
Zeitschrift für vergleichende Physiologie
Vol. 71, 49—128 (1971)

Springer-Verlag Berlin Heidelberg GmbH

IV

Theses

1. De fire anatomiske grupper af sanseceller i vandregræshoppens tympanalorgan har forskellige frekvensfølsomheder.

2. Tympanum indeholder mindst to vibrationssystemer („tynde membran" og hele tympanum), der ved lydpåvirkning foretager svingninger svarende til de rotationssymmetriske svingninger i cirkulære membraner.

3. Centrene for disse to sæt svingninger har forskellige og ikke-konstante placeringer på tympanum.

4. Frekvensdiskrimineringen beror således på rent fysiske forhold: på de to sæt vibrationer og på sansecellegruppernes forskellige anatomiske placeringer på tympanum.

5. Det intakte tympanalorgan fungerer som en trykmodtager ved høje frekvenser.

Contents

VI

Part III: Acoustical properties of the intact ear

Preface

The three papers presented here are the result of some investigations on the physiology and biophysics of the locust ear, which were carried out during the years 1966–70. My interest in these problems was raised by a series of lectures on insect sounds held in 1964 by Dr. S. L. Tuxen at the University of Copenhagen. The results of some preliminary experiments demonstrating the presence of frequency discrimination were published in 1966. At the same time Dr. A. V. Popov, Leningrad, obtained similar results with quite different methods.

The locust ear was the first invertebrate hearing organ demonstrating the ability to discriminate sound frequencies. I therefore decided to investigate the extent of the discrimination and, if possible, the mechanisms involved. The investigations were started at the Zoophysiological Laboratory B, continued (from 1968) at the Zoological Laboratory of the University of Copenhagen, and finished during a year's leave spent at the Lehrstuhl Tierphysiologie, University of Cologne, in 1969–70. A brief note on the findings described in paper I was published in 1968.

The later parts of these studies were concentrated on the physical properties of the ear membrane and its surroundings. Mechanical physics is a hard business for a biologist; but fortunately a number of physicists were willing to introduce the field to me and later point out a number of pitfalls to be avoided. In particular, I should like to thank Mr. O. Juhl Pedersen, M.Sc. and Mr. Knud Rasmussen, M.Sc. for their patience and help. In' addition, several of their colleagues on the staff of the Acoustics Laboratory, The Technical University of Denmark, contributed to the solution of the problems. I am also most indebted to Professor A. Kjerbye Nielsen, to Mr. O. Bengtson, to Dr. Lee Miller, to members of the Laser Research Group at the Royal Institute of Technology in Stockholm, and to the technical staff of the laboratories mentioned above, for suggestions and advice.

These investigations were fairly expensive as compared with most biological studies. It is a great pleasure for me to thank the Carlsberg Foundation, the Deutsche Forschungsgemeinschaft, Statens almindelige Videnskabsfond and Statens naturvidenskabelige Forskningsraad, for their generous support.

During my studies I have reaped the benefit of the pleasant atmosphere and excellent working conditions in the laboratories mentioned. My best thanks are due to the staff and especially to the directors of these laboratories, Professors U. V. Lassen, Bent Christensen and Franz Huber.

I am most grateful to Miss W. Taagerup, to Mrs. H. Holbæk and to Mrs. E. Krenchel for technical and secretarial assistance and to Mr. H. Cowan for correcting the English text. Finally, I should like to thank the Editor of the Zeitschrift für vergl. Physiologie, Professor Dr. Dr. h. c. H. Autrum, both for his careful handling of the manuscripts and for the inspiration I have gained from his pioneering papers on insect hearing.

Dansk Sammenfatning

Introduktion

Markgræshopper har et tympanalorgan på hver side af det første bagkropsled (s. 112, fig. 7a). Det består af en stiv ring af cuticula, hvorover er udspændt en membran, tympanum. Tympanum er bønneformet og ca. 1,5 × 2,5 mm (s. 71, fig. 6). Udefra ses fire fortykkelser af tympanum (s. 71, fig. 6, II). På membranens bagside ses en klump af celler, som fæstner sig til de fire fortykkelser. Hele arrangementet kaldes for Müllers organ. Celleklumpen indeholder fire grupper af sanseceller (a, b, c og d), der via andre celler fæstner til hver sin af de fire fortykkelser. På s. 71 (fig. 6, I) ses Müllers organ set omtrent parallelt med membranen og med de fire grupper af sanseceller indtegnede. Membranen er ret inhomogen; foruden de fire fortykkelser ses en mindre del af membranen at være 3—4 gange tykkere end den øvrige del. Denne ,,tykke membran" er delvis adskilt fra den ,,tynde membran" af de nævnte fire fortykkelser samt af en stavlignende fortykkelse.

I denne afhandling beskrives funktionen af vandregræshoppens øre (tympanalorgan). Det vises, at de fire anatomiske grupper af sanseceller i øret har forskellige frekvensfølsomheder. Dyret skulle derfor have mulighed for at skelne tonehøjder (frekvensdiskriminering). Frekvensdiskrimineringen er baseret på nogle selektive resonanser i trommehinden (tympanum). De fire gruppers frekvensfølsomheder er således bestemt af tympanums anatomi og fysiske egenskaber. Disse resultater blev opnået ved anvendelse af et meget simpelt (,,isoleret") præparat af øret; det bestod blot af trommehinden, sansecellerne og lidt af det omgivende væv. Forskellene mellem dette isolerede præparats og det intakte øres følsomhed lader i det store hele til at være en simpel konsekvens af de forskellige akustiske forhold.

I. del: Sansecellernes frekvensfølsomhed

Den enkelte sansecelles reaktion på lyd kan måles ved at føre en mikroelektrode med en spidsdiameter på ca. 0,2 μm ned i nærheden af cellen og forstærke de nerveimpulser, der affyres af cellen. For at opnå brugbare oplysninger må man sørge for veldefinerede akustiske forhold omkring øret (frit lydfelt). Cellernes svar på en ren tone af kendt frekvens og intensitet kan så optegnes som vist på s. 54—57, fig. 4—10. De fire anatomiske grupper af sanseceller viser sig at have forskellige frekvensfølsomheder (s. 59, fig. 11, viser tærskelkurverne for de fire celle-

grupper). Det er karakteristisk, at følsomheden for en stor del af cellerne
er maximal ved mere end et frekvensbånd (s. 54—57, fig. 4, 8 & 9 viser
eksempler herpå). Svarene fra forskellige celler i samme gruppe (men
fra forskellige præparater) varierer ofte meget (s. 56, fig. 7 viser to
svartyper hos c-cellerne.) Det konkluderes, at de fundne forskelle imellem cellegruppernes frekvensfølsomheder skulle tillade en ret omfattende
frekvensdiskriminering, selv om det isolerede øre, der blev anvendt ved
disse målinger, afviger en del fra det intakte øre (se III. del).

II. del: Frekvensdiskrimineringens fysiske grundlag

I II. del (s. 63—101) beskrives det fysiske grundlag for ørets frekvensdiskriminering. Tympanalmembranens fysiske egenskaber (dimen-
sioner, vægt og elasticitet) er blevet målt. På grundlag heraf beregnes
resonansfrekvenserne for de forventede vibrationer. Endelig er de fak-
tiske vibrationer blevet observeret, dels med laser holografi, dels ved
hjælp af en kapacitiv målemetode. De herved opnåede oplysninger om
membranvibrationerne er dernæst blevet sammenlignet med sansecellernes frekvensfølsomheder. Det konkluderes, at frekvensdiskrimineringen er baseret på rent fysiske fænomener: på sansecellernes forskellige
anatomiske beliggenhed og på tilstedeværelsen af mindst to sæt selektive
resonanser i tympanalmembranen.

Fra teorien for cirkulære membraners vibration skal man forvente
en række forskellige vibrationsmønstre under forskellige betingelser
(s. 65, fig. 1). Hvis membranen er homogen og cirkulær og påvirket af en
jævnt fordelt kraft (lyd), så skal man kun forvente rotationssymmetriske
vibrationer (s. 65, fig. 1, a—c). Dette viser sig omtrent at være tilfældet
her, selvom tympanalmembranen hverken er cirkulær eller homogen.
Hvert svingningsmønster („mode") er en stabil vibrationsmulighed i et
bestemt frekvensområde omkring en resonansfrekvens. Når lydfrekven-
sen øges, afløser vibrationsmønstrene gradvis hinanden, idet knude-
linierne vandrer mod centrum, og nye dannes ved randen.

Den væsentligste afvigelse fra forholdene i den ideale, homogene
membran er, at der i tympanalmembranen findes *to* sæt vibrationer,
der er forårsaget af henholdsvis hele tympanum og af den tynde membran
(se s. 71, fig. 6). Disse vibrationer har forskellig beliggenhed: „tynd-
membran vibrationerne" har deres centrum i eller nær ved centrum for
den tynde membran (s. 81 & 82, fig. 13 & 14). Derimod er „hele-membran
vibrationernes" centrum lokaliseret på eller tæt ved den tykke membran (s. 81 & 83, fig. 13 & 15). Ydermere er beliggenheden af vibrationerne ikke konstant. Tynd-membran-vibrationerne kan således have
deres centrum midt på den tynde membran (2. mode), ved a-cellernes
tilheftningssted (1. mode = grundsvingningen) eller midt imellem disse
positioner (3. mode).

Ved enhver frekvens vil der således være mindst to vibrationer til stede på membranen. Det viser sig, at de to svingningssæt hver især omtrentlig bestemmer vibrationerne inden for deres respektive „territorier". I grænseområdet imellem dem er membranens svingning sandsynligvis bestemt af den vektorielle sum af to vibrationer. Den resulterende vibration afhænger således både af amplitude og fase af de svingninger, der bidrager til vibrationerne i grænseområdet. Denne situation kan for tiden ikke helt overskues. Der er imidlertid foretaget beregninger over de forventede vibrationer for hvert af de to systemer (s. 75—80), under forudsætning af ideelle tilstande, dvs. for en cirkulær, homogen membran. I dette ideale tilfælde er kun centrums vibration nogenlunde enkel (s. 77, fig. 9). Vibrationen af andre dele af den ideale membran kan være betydelig mere kompliceret (s. 78, fig. 10). Hertil kommer, at vibrationscentrenes placering ikke er konstant i tympanalmembranen. Selv om der kun var et sæt vibrationer på membranen, ville det derfor være ret vanskeligt at beregne vibrationen for et enkelt punkt på membranen. De simple eksempler på addition af to vibrationer (s. 79, fig. 12) må derfor blot betragtes som en illustration af princippet.

Med alle disse komplicerende faktorer er det bemærkelsesværdigt, hvor tæt de observerede resonansfrekvenser ligger ved de beregnede. Resonansfrekvenserne ligger ret nær ved sansecellernes foretrukne frekvenser. Det er derfor rimeligt at antage, at en sansecelles frekvensfølsomhed er bestemt af vibrationen af den tympanale fortykkelse, hvortil den fæstner. Denne antagelse har kunnet bekræftes for to af cellegrupperne (s. 90 & 92, fig. 16 & 17). Den anvendte teknik tillader ikke målinger fra de resterende to grupper.

Der er ved disse undersøgelser anvendt stærkt varierende lydintensiteter. Hologrammerne og de kapacitive målinger krævede således 50—60 dB større lydtryk end målingerne fra sansecellerne. Det viser sig imidlertid, at både membranspændingen (s. 72, fig. 7) og gnidningstabene (s. 85—87) er lineære over det anvendte intensitetsområde. Tabene ved gnidning kunne bestemmes på to forskellige måder. De kunne dels beregnes ud fra svar- og resonanskurvernes relative bredde (Q-faktoren), dels fra de absolutte vibrationsamplituder ved en given drivende kraft. Den drivende kraft blev bestemt indirekte i III. del. De således bestemte gnidningstab var omtrent identiske.

III. del: Det intakte øre

I de ovenfor refererede forsøg blev der anvendt et „isoleret" præparat af øret, som blot bestod af trommehinden, Müllers organ og lidt af de omgivende strukturer. Dette præparat var forsøgsteknisk og beregningsmæssigt lettere at håndtere end det intakte øre, men det intakte øre er både anatomisk og fysisk ret forskelligt fra det isolerede præparat.

Det er derfor ikke muligt umiddelbart at anvende resultaterne fra disse forsøg til at beskrive det intakte dyrs hørelse. Når en membran anbringes i væggen af en beholder (her græshoppens krop), må man forvente ændringer både af resonansfrekvenserne, af den drivende kraft og af systemets dæmpning.

For at belyse forskellen mellem det intakte og det isolerede øre er ørets tærskelværdi blevet bestemt ved hjælp af afledninger fra hele tympanalnerven. Man får herved en ide om forskellen mellem absolutte følsomheder; men det må understreges, at denne metode giver langt færre oplysninger end afledninger fra enkelte celler. Det intakte øres tærskelkurve (s. 108, fig. 3) udviser to minima (frekvenser med maximal følsomhed), et ved lave frekvenser (omkring 3,5 kHz) og et ved „høje" frekvenser (omkring 12—16 kHz). Følsomheden ved lave frekvenser varierer dog ikke mindre end 25—30 dB. Denne variation viser sig at være korreleret med den mængde fedtvæv, der kan uddissekeres fra bagkroppen (s. 109, fig. 4 & 5). Følsomheden ved høje frekvenser er imidlertid næsten uafhængig af fedtmængden (s. 108 & 109, fig. 3 & 5). I nogle tilfælde skyldes en ringe følsomhed, at ovarierne er rykket frem til området mellem de to øren. Da det er muligt at forøge følsomheden hos ufølsomme dyr ved at fjerne tarmen, ligger det nær at antage, at følsomheden ved lave frekvenser er bestemt af den mængde væv, der befinder sig i dyret i området mellem de to øren.

Der er nu anvendt en række forskellige præparater til at indkredse de fysiske faktorer, der bestemmer det intakte øres følsomhed (s. 112, fig. 7). Et præparat af øret (det „opererede øre": fig. 7 b) har en kanal ind til tympanums bagside. Dette præparat har en meget konstant tærskelkurve (s. 110, fig. 6), hvis lav-frekvens minimum ligger ca. 4 dB over det mest følsomme intakte øres. Ved fem præparater (s. 112, fig. 7 d—h) er der blevet indopereret en målemikrofon på forskellige steder i dyret. Herved er det muligt at opnå et groft skøn over henholdsvis diffraktionseffekten (s. 113, fig. 8), lydledningen igennem græshoppen (s. 114, fig. 9), og lydabsorbtionen i de indre væv (s. 115, fig. 10). Det viser sig, at græshoppens indre væv fungerer som et akustisk „low-pass" filter, dvs. at lave frekvenser kan passere igennem dyret med minimal dæmpning, medens høje frekvenser dæmpes ret kraftigt.

Dette resultat er i åbenlys modstrid med den hidtil gængse opfattelse af insektøret som trykgradientmodtager. I trykgradientmodtagere påvirker lydbølgen både for- og bagsiden af membranen (s. 105, fig. 2 b). En sådan lydmodtager vil være særdeles følsom for lydbølgens retning, idet den drivende kraft teoretisk vil være lig nul, når lydbølgens udbredelsesretning er parallel med membranen. Det har været antaget (Pumphrey, 1940; Autrum, 1941), at insekter udnytter dette forhold ved retningsbestemmelse af en lydkilde. Da høje frekvenser absorberes af de indre væv, må græshoppens øre fungere som en trykmodtager i det frekvensområde, der registreres af d-cellerne (se ovenfor). Ved retnings-bestemmelse af højfrekvente lydkilder er dyret derfor henvist til at udnytte de forskelle i lydtryk, der skyldes diffraktion. Det er uvist, især i fede dyr, i hvilket omfang øret fungerer som trykgradientmodtager ved lave frekvenser. I magre dyr lader forholdene ved lave frekvenser sig tilnærmelsesvis beskrive ved en model af en asymmetrisk lydmodtager (s. 105, fig. 2 c), hvor man må tage hensyn til strålingsimpedanserne fra for- og bagsiden af membranen (s. 117 & 119) og til effekten af det lukkede hulrum bag øret (s. 117).

Disse beregninger fører frem til en omtrentlig bestemmelse af den drivende kraft i det opererede øre (s. 118). Herfra kan man ved sammen-ligning af følsomhederne anslå den drivende kraft i det isolerede øre (s. 120, fig. 12).

Det må imidlertid understreges, at den anvendte model er alt for simpel til at tillade en egentlig analyse af det intakte øre. Man skulle således forvente en langt kraftigere forskydning i grundresonansfrekven-serne end den, der er observeret. Afvigelserne fra det forventede kan skyldes, at græshoppens væv er akustisk „bløde". Der foreslås nogle metoder til yderligere undersøgelser af disse forhold.

III. del afsluttes med et appendix, hvori de hyppigst anvendte meto-der til bestemmelse af tærskelværdier bliver sammenlignet. Det har desværre vist sig umuligt at sammenholde de målinger, der er beskrevet i dette arbejde, med andre undersøgeres resultater. Årsagen hertil er dels de mangelfulde akustiske forhold, de fleste undersøgere har arbejdet under. En væsentlig hindring er imidlertid de usikre definitioner på tærskelværdier, der har været anvendt. På side 127, fig. 13, er således vist tærskelkurven for *Locusta migratoria*. Omtrent halvdelen af den viste forskel skyldes den anvendte metode, medens den resterende halv-del må føres tilbage til variationer i størrelsen af dyrenes fedtlegemer.

Konklusion

Vandregræshoppens øre er det første eksempel på en lydmodtager hos hvirvelløse dyr, der muliggør en skelnen af tonehøjder. Det er des-uden det første eksempel på, at der i naturen er realiseret høreorganer, hvor tonehøjder skelnes ved hjælp af selektive resonanser. Dette princip ligger til grund for den bekendte Helmholtz'ske teori for hvirveldyrørets funktion. Det må dog understreges, at græshoppeøret både anatomisk og fysisk adskiller sig væsentligt fra hvirveldyrøret. Desuden er både grund- og oversvingninger nyttiggjort til frekvensdiskriminering i græs-hoppens øre, medens Helmholtz kun opererede med grundsvingninger.

På trods af disse forskelle mellem græshoppeøret og hvirveldyrøret synes tympanalorganet dog velegnet som et simpelt „modelsystem" ved

XIV

fremtidige undersøgelser af basale høremekanismer. Dette synes navnlig at være hensigtsmæssigt ved to sammenkoblede typer af problemer. Den ene type vedrører sammenhængen imellem tids- og frekvensopløsningerne i høreorganer, hvor den primære lydanalyse er baseret på mekaniske systemer (se s. 87). Det andet problemkompleks vedrører den grad af selektivitet, der kan opnås i sådanne systemer; dvs. hvor meget af lydanalysen der kan udføres mekanisk, og hvor meget der må overlades til nervøs bearbejdning.

Disse fysiologiske og biofysiske problemer er også direkte relevante for forståelsen af lydmodtagernes (begrænsende) rolle i overførelsen af information med akustiske signaler. Vi ved for tiden ikke, hvortil græshoppen bruger sin evne til frekvensdiskriminering. Man ved heller ikke, hvad der er de væsentligste informationsbærende parametre i lydsignalerne. De hidtil undersøgte øren hos insekter viser tilsyneladende en vid variation i udvælgelsen af de betydningsfulde parametre. Frekvens, tidsopløsning, intensitet og rytme er de parametre, der varierer i insekternes sang. En forståelse af dette problemkompleks synes mest hensigtsmæssigt at kunne opnås, hvis man gør sig de begrænsninger klart, der hidrører fra anvendelsen af mekaniske strukturer ved lydproduktion og lydmodtagelse. Græshoppens akustiske kommunikation synes velegnet til en sådan analyse.

Z. vergl. Physiologie 71, 49—62 (1971)
© by Springer-Verlag 1971

The Physiology of the Locust Ear

I. Frequency Sensitivity of Single Cells in the Isolated Ear

Axel Michelsen

Laboratories of Zoology and Zoophysiology B
University of Copenhagen

Received November 3, 1969 *

Summary. The sensory responses of single receptor cells in the isolated ear of the locust *Schistocerca gregaria* were measured under controlled acoustical conditions. The four anatomical groups (Fig. 1) differ as to frequency sensitivity (Fig. 11). Although the isolated ear differs much from the intact ear, it may be concluded that fairly accurate information about sound frequency reaches the CNS. The responses of most units showed a maximum sensitivity at two (Figs. 4 and 9) or three (Fig. 8) different frequencies. But several units had only one maximum (Figs. 6 and 7, right).

Introduction

Information about the frequency of a sound may be signalled by auditory receptors in two different ways. The anatomical groups of receptor cells may differ as to characteristic frequency (CF, i.e. the frequency of maximum sensitivity), thus providing the CNS with information about frequency (the place principle). At low frequencies the receptor cells may also send trains of nerve impulses to the CNS, each corresponding to a certain phase of the sound wave (the telephone principle). In the ear of vertebrates (see Whitfield, 1967), the place principle is the most likely explanation for frequency discrimination in the entire frequency range of the ear, and the telephone principle is normally thought to be of minor importance, and that only at frequencies below 2–4 kHz. So far, no examples of frequency discrimination based upon the place principle have been found outside the vertebrates.

In invertebrates the telephone principle is illustrated by the response of cercal hairs of insects to low pitched sounds (Pumphrey and Rawdon-Smith, 1936). It is doubtful, however, whether the insect CNS extracts the information about frequency carried by the cercal nerve.

Insects have generally been considered tone-deaf; the response of the insect ear was thought to be determined by changes in amplitude of the sound, and the frequency (pitch) was not considered important. In 1960,

---

* Editor's note: In agreement with the author the publication was delayed until parts II and III were ready.

however, Horridge obtained evidence for some degree of pitch discrimination in the tympanal organ of locusts by recording from the tympanal nerve and central connectives. Popov (1965), by means of selective destruction, and Michelsen (1966), by recording from the individual receptor cells found evidence for two groups of receptor cells with widely different characteristic frequencies, and apparently with little overlap of the frequency responses at moderate intensities.

The present paper describes different frequency responses of the four anatomical groups of receptor cells in the *isolated* tympanal organ of the desert locust. The relationship between the results on the isolated ear and those obtained in the ear *in situ* will be dealt with in a separate publication (paper III). The amount of overlap of some of the frequency responses seems to allow a rather accurate determination of frequency in the CNS.

Material and Methods

The locust has a tympanal organ on each side of the first abdominal segment. It consists of a sclerotized ring forming a recess in the abdomen and encircling a membrane, the tympanum. The tympanum is bean-shaped and about 2.5×1.5 mm at its widest. The sensory units, a total of 60–80 chordotonal sensilla, are attached in four groups to four modified parts of the tympanum (Fig. 1). The attachment parts of the drum are thickened regions which form an *elevated process* (an invagination about 100 μm deep), a *styliform body*, a *folded body*, and a *pyriform vesicle*. The groups of sensory cells attached to these regions were named the a-, b-, c-, and d-groups respectively by Gray (1960), who described the fine structure of the ear, and this terminology will be used here. The receptor cells of the a-, c- and d-groups are orientated in three almost mutually perpendicular planes. The fourth group (b) is orientated in the same plane as the a-group. The a-group comprises about half of the total number of receptor cells, and the three other groups about 8–12 cells each.

The female locusts (*Schistocerca gregaria* Forskål, ph. gregaria) used were supplied by the Anti-Locust Research Centre, London, and were about 4 weeks after the final moult. The sclerotized ring encircling the tympanum and a small portion of the surrounding tissue were removed and mounted on a small platform of wax. When mounted the tympanum was about 15° askew relative to the vertical plane (see Michelsen, 1966).

The platform with the preparation was placed on the end of a vertical brass rod in an open box made of "soft masonite" and lined on the inside with 15 cm of mineral wool (Fig. 2). The box was situated in an isolation room, which had a special sound attenuating steel door.

The two loudspeakers were situated at the hindmost wall of the box. A wall of mineral wool was placed about one meter from the opening of the box. Since this wall was somewhat askew relative to the box (see Fig. 2), very little echo reached the preparation. The micromanipulator was about 20 cm from the preparation and hidden behind 5 cm of mineral wool. The microscope could be swung away during the experiment. In this way a good approximation to an acoustic free field was obtained: the sound pressure was almost constant around the preparation, and the echo (measured with a microphone pointing towards the sound source) was at least 30 dB below the signal. In the later experiments the echo was

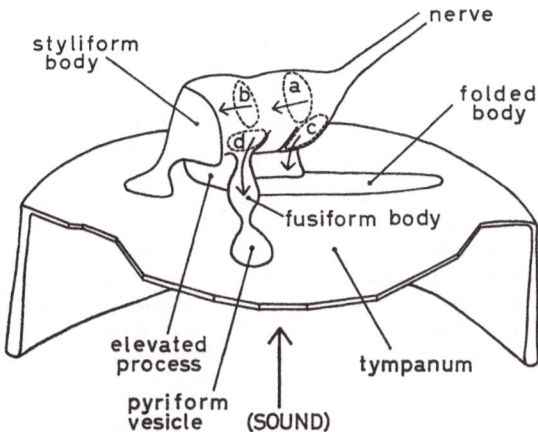

Fig. 1. The anatomy of the left tympanal organ (redrawn after Gray, 1960). The letters *a–d* indicate the position of the four groups of receptor cells. The arrows indicate the direction of the dendrites

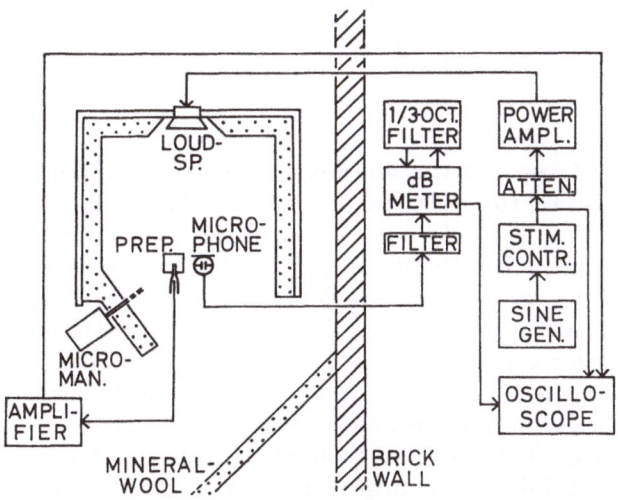

Fig. 2. The arrangement of the apparatus and sound absorbing material (cf. the text)

further reduced by means of cone-shaped Moltopren microwave absorbers (Grünzweig & Hartmann, type P100) placed behind the preparation.

The distance from the loudspeakers to the microphone and preparation was about 70 cm. At this distance and at the frequencies used here the sound reaching the ear and microphone could be regarded as plane waves. In a plane wave the pressure gradient (the sound parameter acting to move the *isolated* tympanum) can be estimated directly from the sound level measured by means of a condenser microphone.

4*

The sound generating and sound measuring apparatus was essentially identical with that previously described (Michelsen, 1966), but with the sine wave generator, amplifiers, and loudspeakers replaced by selected low-distortion types. The sound levels are indicated in dB root mean square relative to $2 \cdot 10^{-5}$ Newton/m² (about human threshold at 1 kHz). The background noise level at the preparation was below 25 dB root mean square in the frequency range 0.5–20 kHz. In the present experiments, sound pulses of 100 msec duration, with a rise and decay time of 2 msec were used. The sine wave started and ended at zero, and no "clicks" were produced at the beginning and end of the signals.

When the threshold is determined for units having a sharp low-frequency cut-off, the amount of harmonic distortion of the signal becomes critical at frequencies below the CF. The harmonics were measured by means of an $^1/_3$ octave filter (Brüel & Kjär) and were at least 30 dB below a signal of 90 dB sound level. Although at most frequencies the harmonics were more than 40 dB down, they still presented a problem in some of the recordings.

Extracellular action potentials were recorded by means of glass capillary microelectrodes (tip diameter 0.2 μm, impedance 20–40 Mohm, filled with 1.5 Molar NaCl). When the sensitivity of a cell to a reference sound pulse changed more than ± 2 dB during the course of the experiment, the cell was discarded.

In the absence of sound the receptor cells showed "spontaneous" activity, typically about 1–10 spikes/sec. Since spontaneously active cells do not have a well defined threshold, the "threshold" is defined here as the intensity necessary to give an average response of one spike more than the spontaneous activity to a sound pulse of 100 msec duration.

Results

A series of responses to sound stimuli were recorded from 28 individual a-cells, 20 b-cells, 16 c-cells, and 18 d-cells. In about half of the recordings the cell met the stability criteria mentioned above, thus permitting a determination of the threshold curve. In a few cases the frequency response of two receptor cells was determined from the same preparation, but most of the recordings were from different preparations. The properties of the receptor cells varied considerably within each of the four anatomical groups. However, the difference in frequency response between the groups appeared to be highly significant.

In a large number of cells there was a tendency for a second (and sometimes a third) maximum sensitivity at a higher frequency than the CF. These "resonances" in the response cannot be explained as artefacts due to errors in the calibration of the sound sources, since in several cells the "resonance" was very small or entirely lacking. In a few cells, on the other hand, the response at the second maximum was as large or even larger than the response to the "CF" itself (see below).

The a- and b-Cells

The anatomical position and frequency response of the b-cells are so close to those of the a-cells that for some time the recordings from the

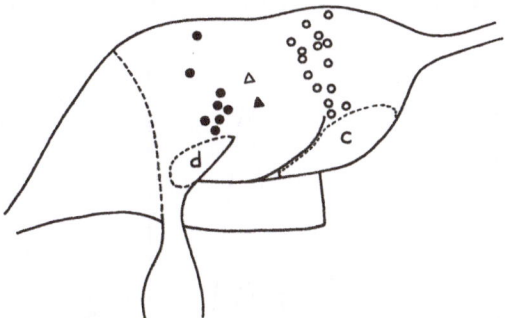

Fig. 3. The anatomical position of the recordings from a- and b-cells (○ a-cell,
● b-cell, △ cell with a-cell properties, ▲ cell with b-cell properties)

Table. *Data for 15 a-cells and 8 b-cells. Further explanation in the text*

	a		b		P
	$\bar{x}$	$S\bar{x}$	$\bar{x}$	$S\bar{x}$	
Characteristic frequency (Hz)	3747	64	3463	98	0.02
Threshold at CF (dB)	46	0.9	51	1.3	<0.01
SDB (dB/kHz)	7.8	0.6	16.1	1.8	<0.001
SDA (dB/kHz)	7.6	0.6	13.4	1.5	<0.001
Intensity-response slope (spikes/10 dB)	12.1	0.6	15.0	0.7	<0.01

two groups were not considered different. There are, however, a number
of statistically significant differences between the properties of the two
groups. In Fig. 3 the anatomical position of the recordings from 15 a-cells
and 8 b-cells are shown, and some of their properties are compared in the
table. The Table indicates the mean values ($\bar{x}$), the standard error of the
mean ($S\bar{x}$), and the probability (P) that the groups are identical (t-test).

The average threshold at CF was calculated for the most sensitive
$2/_3$ of the cells in each group (10 a-cells and 6 b-cells). The sensitivity
decrease for frequencies below (SDB) or above (SDA) the CF are given
in dB per kHz. These two values taken together indicate the degree of
tuning to the CF, but as seen on Fig. 5 they are not always of the same
magnitude. The intensity-response curve is sigmoid, and therefore the
"intensity-response slope" is calculated from the increase in sound
level needed at CF to obtain an increase from 5 to 15 in the number
of spikes per 100 msec. Smaller values for the intensity-response slope
were found on both sides of the CF.

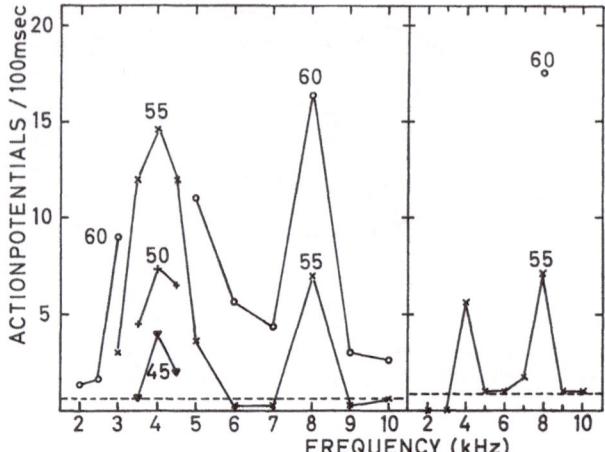

Fig. 4. The responses of two a-cells to sound pulses of 100 msec duration at varying frequencies. Numbers by each curve indicate the sound level in dB (rel. $2 \cdot 10^{-5}$ N/m²). Broken lines show the average level of spontaneous activity. Note the second sensitivity maximum at 8 kHz

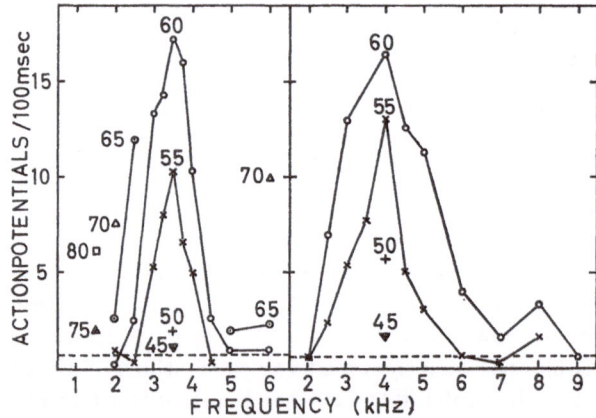

Fig. 5. The responses of two a-cells (for explanation see Fig. 4). Note the small second maximum in the right graph

For all parameters except the characteristic frequency, the difference between the two groups is highly significant. Thus, the b-cells are about 5 dB less sensitive than the a-cells, and they are more sharply tuned to their CF, which may be about 0.3 kHz lower than that of the a-cells.

In about half of the a-cells investigated a second response maximum was found around 8 kHz (Fig. 4). In a few cells this second maximum

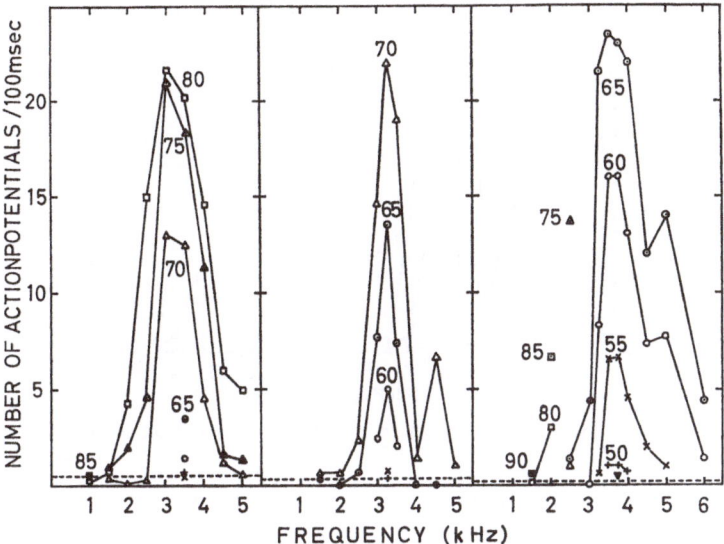

Fig. 6. The responses of three b-cells. Explanation see Fig. 4. Note the very sharp tuning of the cell in the middle graph and a second sensitivity maximum at 5 kHz in the right graph

was found at 6 or 7 kHz. The sensitivity at 8 kHz was on the average 4 dB lower than that at the CF (3.5–4 kHz), but two cells responded more vigorously at 8 kHz than at their "CF" (Fig. 4). In about one third of the a-cells, however, a second response maximum was reduced or absent (Fig. 5).

In the b-cells (Fig. 6) about half of the units had a small second maximum (5–10 dB down) at about 5 kHz.

The patterns of adaptation of the two groups are very similar to each other. The spike frequency is a maximum during the first 50 msec after the onset of sound and decreases during the next 200–300 msec to reach a slowly decreasing plateau (see Michelsen, 1966).

The c-Cells

The frequency responses obtained from the individual c-cells may be divided into two groups. In some c-cells the sensitivity was rather uniform within the frequency range 2–3.5 kHz, and outside this range the sensitivity decreased markedly (Fig. 7, right). In most c-cells, however, the sensitivity at 1.5 kHz was as high or even higher than that at 2.5–3 kHz (Fig. 7, left). In such cells the sensitivity to 2 kHz

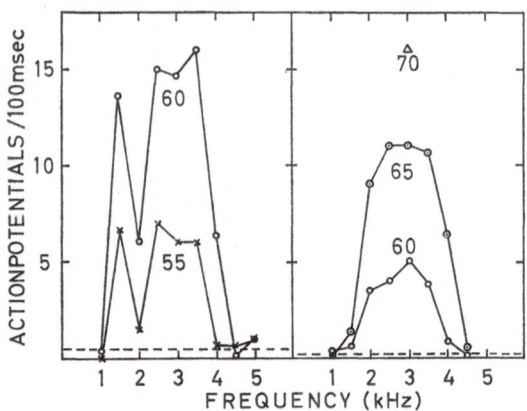

Fig. 7. Two response types in c-cells (for explanation see Fig. 4)

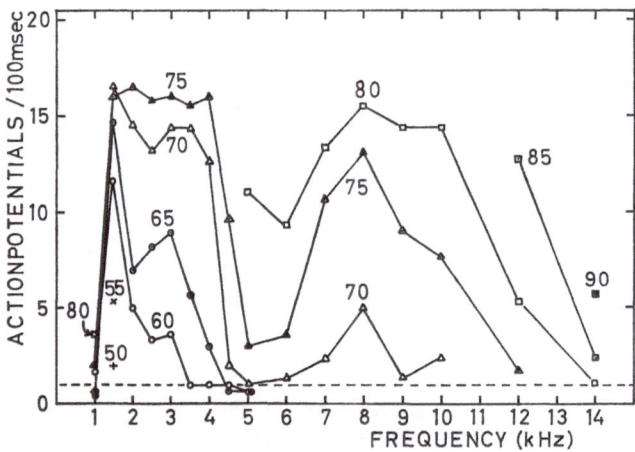

Fig. 8. The responses of a c-cell having three sensitivity maxima (for explanation see Fig. 4). Note the 28 dB difference in sensitivity to 1 kHz and to 1.5 kHz

was about 5 dB below that at 1.5 kHz. In some units a small third maximum of sensitivity was found at 8 kHz (Fig. 8).

The pattern of adaptation of the c-cells is rather different from that of the other groups. The spike frequency increases more slowly, to reach a maximum 100–400 msec after the onset of sound. After the initial rise the spike frequency decreases slowly in a phasic manner. The slow response of the c-cells, in contrast to the fast response of the other groups, is also evident by their failure to respond to single "clicks" at high repetition frequencies (Michelsen, 1966).

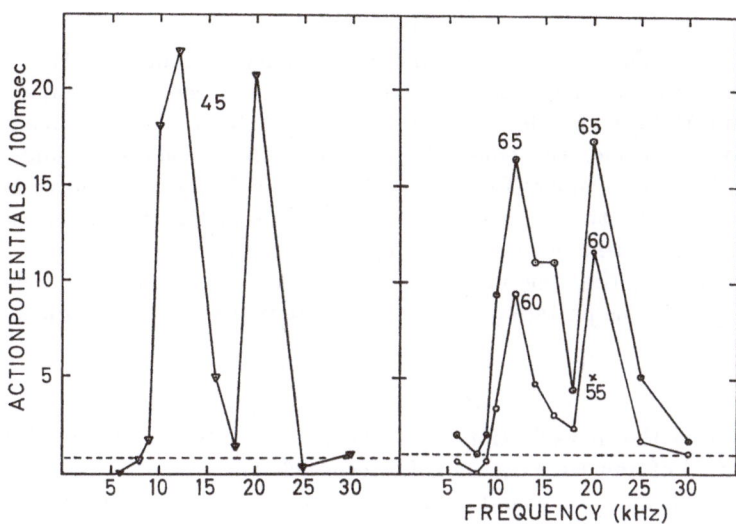

Fig. 9. The responses of d-cells having two sensitivity maxima (for explanation see Fig. 4)

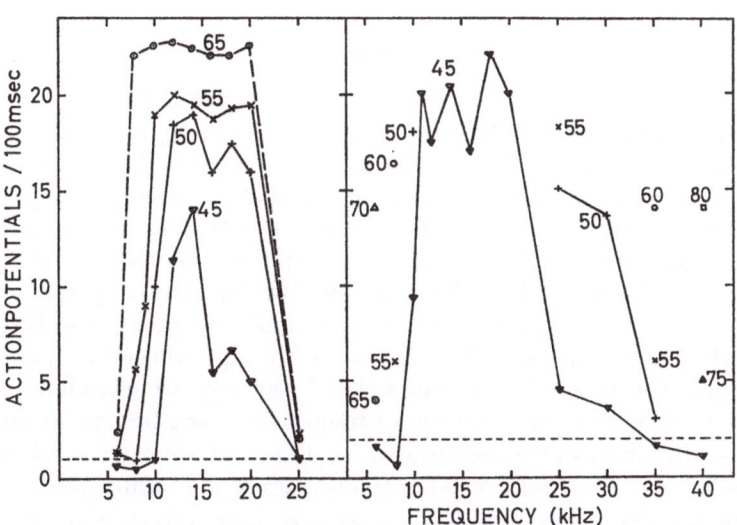

Fig. 10. The responses of d-cells (for explanation see Fig. 4). Left: a cell with a small second sensitivity maximum (unbroken line), and the response of another cell (long, broken lines). Note the large decrease in sensitivity towards low frequencies in both graphs

The d-Cells

The d-cells have their maximum sensitivity at about 12 kHz (range 10–14 kHz), and most cells also have a second sensitivity maximum around 19 kHz (Fig. 9). The sensitivities at these two maxima were almost the same. In some cells, however, the second maximum was much less conspicuous (Fig. 10). At 6 and 40 kHz (i.e. one octave below and above the two sensitivity maxima) the sensitivity is about 35 dB less. A remarkable property of many d-cells is their very limited "measuring range" (i.e. the intensity range from threshold to saturation, see Fig. 10). The pattern of adaptation is equivalent to that of the a- and b-cells.

Low-frequency Units

During the investigation of d- and b-cells, action potentials from a fifth kind of unit were occasionally recorded. The spikes, as judged from their small amplitude and negative polarity, were probably generated by axons of small diameter. These units responded to sound of high intensity (90 dB) and low frequency (some hundred Hz). In two cases it was possible to record similar responses from units in the fusiform body (see Fig. 1). Probably, these responses originate from mechanoreceptors situated upon the tympanal membrane near to the pyriform vesicle.

Discussion

Frequency Discrimination and the Intact Ear

These studies show that the maximum sensitivity of each group of receptor cells is limited to a small number of discrete frequency bands. For most cells the frequencies of maximum sensitivity were around 3.7 kHz for the a-cells, 3.5 kHz for the b-cells, and 12 kHz for the d-cells. In some cells a second sensitivity maximum was observed around 8 kHz (a), 5 kHz (b), and 19 kHz (d). The response of the c-cells is more complex, but the preferred frequencies are 1.5 kHz, 2–3 kHz (all cells), and 8 kHz. In the following publication (paper II) the physical properties of the isolated ear will be considered with special reference to the problem of frequency discrimination. It will be shown that most of the frequencies mentioned above correspond to the expected and observed resonance frequencies of the tympanum.

In the present experiments the sensory units attached to different anatomical parts of the tympanum showed different frequency sensitivities. In these experiments the ear was completely *isolated*, i.e. the preparation merely consisted of the sclerotized ring, tympanum, and receptor cells. Although the isolated ear, considered as a mechanical

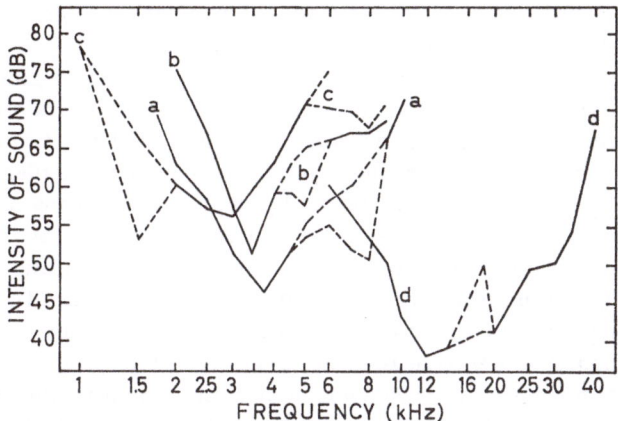

Fig. 11. The threshold curves for the four groups of receptor cells in the isolated locust ear. Broken lines indicate variations in threshold curves for different cells within each group

system, is far from simple, the acoustic properties of the *intact* ear are much more complicated. The intact pair of ears are connected by a series of air sacs formed by tracheal membranes. In a following publication (paper III) it will be shown, how the presence of these air sacs may affect the sensory response of the ear to sound stimuli.

Because of the physical difference between the isolated and intact ears, the present experimental results do not tell us how much information about frequency is actually sent to the CNS by the receptor cells in the intact ear. In paper III it will be shown that some of the main preferred response frequencies of the isolated ear are also found in recordings from intact ears. Therefore on the basis of the threshold curves of the isolated ear, one may speculate about the ability of the intact insect to carry out a pitch discrimination. According to Haskell (1957) the flight noises of locusts are in the intensity range 50 to 65 dB and around 4 kHz. It has been suggested (Michelsen, 1968) that in this range locusts might be able to perform an approximate determination of frequency be means of very few central neurons: a shift of frequency at an appropriate intensity causes different groups of receptor cells to respond in turn (Fig. 11). At low frequencies the c-cells alone respond, and at higher frequencies the combinations are: ac, abc, ab, a, ad, and d. If this primitive type of frequency discrimination is used, the functional significance of the b-cells becomes comprehensible. Assuming this kind of central mechanism exists, the response of the b-cells would be useful, even if the CF's of the a- and b-cells were identical. If a larger number of central neurons were used, the pitch discrimination (together with the intensity discrimination) might improve considerably.

It is known that the information about frequency signalled from the intact ears is transmitted to the brain. In *Locusta migratoria*, Popov (1967) found two ascending neurons on each side of the third thoracic ganglion. The responses of these neurons are very complex, and information about frequency seems to be signalled by means of different types of adaptation; low (3–8 kHz) and high (12–20 kHz) frequencies evoke a phasic and tonic response, respectively. According to Adam and Schwartzkopff (1967) the same information is found in the response of neuron groups in the protocerebrum of *Locusta*. They found three types of neurons, sensitive to 4–8, 12–30, and 4–30 kHz, respectively. Thus, it is known that at least some information about frequency reaches the brain. The results of Yanagisawa, Hashimoto and Katsuki (1967) suggest, however, that the response patterns of higher neurons may be markedly influenced by inhibitory interactions between various auditory inputs. The studies of Adam (1969) on the behaviour of auditory neurons in the brain likewise suggest that the processing of auditory information in the CNS is extremely complex. Unfortunately, most authors do not indicate the exact acoustical condition of their experiments, and therefore their results are difficult to compare.

Sources of Error

In the present experiments the microelectrode was inserted into the Müller organ. Although the electrodes used were long and thin, their presence in the organ may have influenced the vibration of the membrane and the sensory response to sound. In a few cases it was possible to record from single fibers in the acoustic nerve, and these recordings did not differ from those obtained from the Müller organ. Such recordings were seldom successful, because the sheet covering the nerve hindered the penetration of the microelectrode. Recordings from entire nerves at some mm distance from semi-isolated ears gave threshold curves very similar to those expected from the present results.

Anomalous Results

The results presented above have been selected from the results of all the experiments on the basis of two criteria: First, the cell should meet the criterion for stability mentioned above, and secondly the sensitivity and frequency response should be reasonable "normal". For example, the threshold at CF for the most sensitive $^2/_3$ of the a-cells was about 46 dB (range 42–50 dB). In some cases and previously (Michelsen, 1966), thresholds of about 60 dB and frequency ranges of 2–10 kHz have been found for a-cells, and these less sensitive cells also tend to have a higher CF. Typically, a-cells with a threshold of 60–70 dB may have a CF of 5–8 kHz, as compared with an average CF of about 3.7 kHz for the most sensitive a-cells. Similarly, d-cells with a CF of about 25 kHz have occasionally been observed. It will be shown in the following publication that this shift in CF was probably due to a change in the membrane compliance caused by the drying of the preparation.

Recordings from Damaged c-Cells

The c-cells seem to be rather sensitive to disturbances during the recording of single unit activity. I have previously described (1966) the inhibition of spontaneous activity in c-cells by stimulation. I also described the response of an "e-unit". These types of response were probably recorded from damaged c-cells:

In one experiment spikes from two c-cells were recorded simultaneously Initially, one of the cells was excited by low intensity sound (60–70 dB, 3 kHz) and inhibited by intense sound (80–90 dB), whereas the other cell was excited by all the intensities mentioned. Later, the first cell was inhibited at all sound levels above 60 dB, whereas the behaviour of the other cell remained constant.

In another experiment it was possible to follow the gradual change of the response pattern of a cell from the "e-unit" type to that of a typical c-cell. The small spikes, which often initiate the spikes in "e-units" (Michelsen, 1966, Fig. 11), have now been observed in all four groups of receptor cells. They seem most likely to occur in recordings near to the dendrite of the receptor cell; their nature is unknown.

I am grateful to Professors F. Huber and U. V. Lassen for discussions and advice, to Mr. S. Olesen-Larsen and Mr. O. Bengtson for advice on the statistical and technical problems, and to Miss W. Taagerup for able technical assistance. The work was supported by the Anti-Locust Research Centre, London, the Carlsberg Foundation, and Statens almindelige Videnskabsfond.

References

Adam, L.-J.: Neurophysiologie des Hörens und Bioakustik einer Feldheuschrecke (*Locusta migratoria*). Z. vergl. Physiol. **63**, 227–289 (1969).

— Schwartzkopff, J.: Getrennte nervöse Representation für verschiedene Tonbereiche im Protocerebrum von *Locusta migratoria*. Z. vergl. Physiol. **54**, 246–255 (1967).

Gray, E. G.: The fine structure of the insect ear. Phil. Trans. B **243**, 75–94 (1960).

Haskell, P. T.: The influence of flight noise on behaviour in the desert locust *Schistocerca gregaria* (Forsk.). J. Insect Physiol. **1**, 52–75 (1957).

Horridge, G. A.: Pitch discrimination in *Orthoptera* demonstrated by responses of central auditory neurones. Nature (Lond.) **185**, 623–624 (1960).

— Pitch discrimination in locusts. Proc. roy. Soc. B **155**, 218–231 (1961).

Michelsen, A.: Pitch discrimination in the locust ear: observations on single sense cells. J. Insect Physiol. **12**, 1119–1131 (1966).

— Frequency discrimination in the locust ear by means of four groups of receptor cells. Nature (Lond.) **220**, 585–586 (1968).

Popov, A. V.: Electrophysiological studies on peripheral auditory neurons in the locust (In Russian, with an English summary). J. evol. Biochem. Physiol. **1**, 239–250 (1965).

— Synaptic transmission at the level of the first synapses of the auditory system in *Locusta migratoria* (In Russian, with an English summary). In: Evolutionary neurophysiology and neurochemistry (ed. by Kreps, E. M.). Leningrad: Nauka 1967.

— The characteristics of the activity of central auditory neurons in locusts (In Russian). In: Hearing mechanisms. Leningrad: Nauka 1967.

Pumphrey, R. J., Rawdon-Smith, A. F.: Hearing in insects: The nature of the response of certain receptors to auditory stimuli. Proc. roy. Soc. B **121**, 18–27 (1936).

Whitfield, I. C.: Coding in the auditory nervous system. Nature (Lond.) **213**, 756–759 (1967).

Yanagisawa, K., Hashimoto, T., Katsuki, Y.: Frequency discrimination in the central nerve cords of locusts. J. Insect Physiol. **13**, 635–643 (1967).

Axel Michelsen
Zoological Laboratory
Universitetsparken 15
DK-2100 Copenhagen 0, Denmark

Z. vergl. Physiologie 71, 63–101 (1971)
© by Springer-Verlag 1971

The Physiology of the Locust Ear

II. Frequency Discrimination Based upon Resonances in the Tympanum

Axel Michelsen

Laboratories of Zoology and Zoophysiology B, University of Copenhagen, and
Zoologisches Institut, Lehrstuhl für Tierphysiologie, Universität zu Köln

Received September 20, 1970

Summary. 1. The expected resonance frequencies of the tympanal membrane have been calculated from its dimensions, mass, and compliance. The thin part of the tympanal membrane may vibrate independently of the entire tympanum. Thus, there are at least two sets of resonances (Fig. 8).

2. The two sets of vibrations have been observed by means of laser holography (Figs. 13–15) and measured with a capacitance electrode (Figs. 16–18). The position and amplitude of the vibration patterns, the phase relationships, and the inter-action of the two sets of vibration have been studied. The results are compared with the frequency sensitivity of the four groups of receptor cells.

3. The groups of receptor cells are attached to four specialized areas on the tympanum (Fig. 6). The vibrations of these areas of attachment are a maximum at the frequencies of maximum sensitivity in the receptor cells (Figs. 16 and 17). Thus, the frequency discrimination seems to be a purely physical phenomenon, based partly on the presence of the tympanal resonances, and partly on the different positions of the receptor cells on the tympanal membrane.

4. The two sets of vibrations have different spatial positions on the tympanum. The centre of the entire-membrane-vibrations is situated in one end of the membrane (Fig. 15), whereas that of the thin-membrane-vibrations is almost at the centre of the tympanum (Fig. 14). The positions of the centers of vibration are, however, not constant (Figs. 13 and 14). Different modes may have somewhat different centre positions, and these positions may change with frequency because of inter-actions between the two sets of resonances. Therefore, receptor cells attached to different areas on the membrane may pick up different modes of vibration. Also, the receptor cells may almost fail to respond to some modes, if their area of attach-ment is at a nodal circle of these modes at resonance.

Introduction

Many invertebrates can hear (review: Autrum, 1963), but very few can discriminate frequencies. In the locust ear four anatomical groups of receptor cells are attached to four specialized areas on the tympanal membrane. The frequency sensitivities of these groups are different in the *isolated ear*, i.e. a preparation consisting of the tympanum and its cuticular rim (paper I: Michelsen, 1971). In the present paper it is shown that the frequencies of maximum sensitivity in the receptor cells cor-respond to the expected and observed resonances of the normal modes of vibration of the tympanal membrane. It is argued that the frequency discrimination of the locust ear is a purely physical phenomenon, based partly on the presence of the tympanal resonances, and partly on the different position of the receptor cells on the tympanum. In the following publication (paper III) the physics of the *intact* locust ear will be considered. In these papers the SI (meter-kilogram-second) system of physical units will be used.

1. Resonance and Hearing

The presence of resonance in hearing organs is not a new idea (see Whitfield, 1967). Some of the early theories (17th century) postulated the existence of mechanical resonators in the vertebrate ear; the function of these hypothetical resonators was considered to be that of magnifying a weak sound stimulus, by storing incoming sound energy over a number of cycles. Later, the idea was introduced that the resonators might have different resonance frequencies.

These thoughts found their most famous expression in the theory of Helmholtz (1863). He suggested the existence of a resonator in the cochlea, supplied by specific nerve fibers, for each subjectively distinguishable tone. The Helmholtz-theory was later abandoned, partly because of the failure to identify the resonators anatomically. Also, this and other resonance theories failed to explain the fact that in the vertebrate ear a sharp frequency discrimination is carried out together with a fine time resolution. This difficulty could, however, be overcome in the "travelling-wave theory" proposed by Békésy (review: 1960).

Recently, Huxley (1969) has revived the idea of auditory analysis by resonance. He showed mathematically that truly resonant oscillations, the position of which shifts with frequency, may after all be a possibility in the cochlea. The reason for Békésy's observation of travelling waves might, according to this view, be the artifical mechanical conditions created during the experiments. This argument seems to be valid for most experimental studies of hearing organs. The theory proposed by Huxley has been criticized by Békésy (1969, 1970).

Although the possible existence of true resonances in the cochlea is still discussed, no clear-cut example of frequency discrimination based upon the existence of selective resonances has been found so far in a hearing organ. Therefore, from a comparative and theoretical point of view the locust ear is a unique physiological preparation, which allows us to study how selective resonances in an inhomogeneous membrane can be used as the physical basis for a frequency discrimination.

It should, however, be emphasized that the anatomy and the physical properties of the locust ear are extremely different from those of the vertebrate ear. In particular, it should be born in mind that a strong

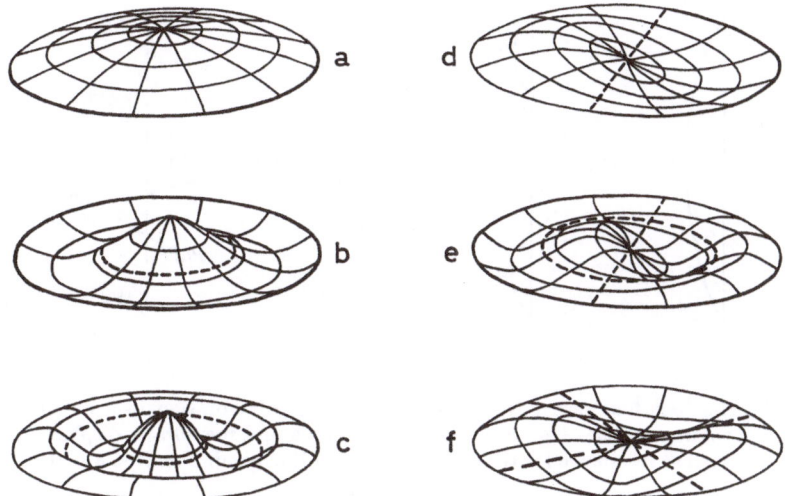

Fig. 1a–f. The first three circularly symmetrical modes of vibration in a circular membrane (a–c), and some other possible modes (d–f). The nodal lines are indicated by dotted lines. Further explanation in the text. (Redrawn from P. M. Morse: Vibration and sound. Copyright 1936, 1948 by the McGraw-Hill Book Company, Inc. Used with permission of McGraw-Hill Book Company)

mechanical coupling probably exists between the basilar membrane and the surrounding fluid, whereas in the locust ear the vibrations of the membrane are hardly influenced by the surrounding medium.

2. The Vibration of Membranes

Descriptions of the vibration of membranes can be found in most textbooks on acoustics, but for the convenience of the reader a few relevant points will be summarized. We shall consider a homogeneous membrane, fastened along a boundary circle, and acted upon by a uniform, harmonic driving force (e.g. a sound wave). The travelling velocity of transverse waves in the membrane is supposed to be much smaller than the velocity of sound waves in the surrounding medium.

At low frequencies the membrane will tend to move as a whole (Fig. 1a), and its behaviour very much resembles that of a simple driven oscillator. At very low frequencies the maximum displacement of the membrane will be almost in phase with the driving force, but as the frequency approaches the first resonance, the displacement will be delayed relative to the force. At the first resonance frequency (f_{01}) this delay is 90° (Fig. 2, 1), and somewhat above the first resonance the instantaneous displacement of the membrane is nearly opposed to the driving force (i.e. the phase lag is about 180°).

If the frequency is increased, a circular nodal line appears at the edge and moves towards the centre. The part of the membrane around the centre has remained out of phase with the force, but the part outside the nodal circle is almost in phase with the force. The amplitude of the overall motion now increases, and a second

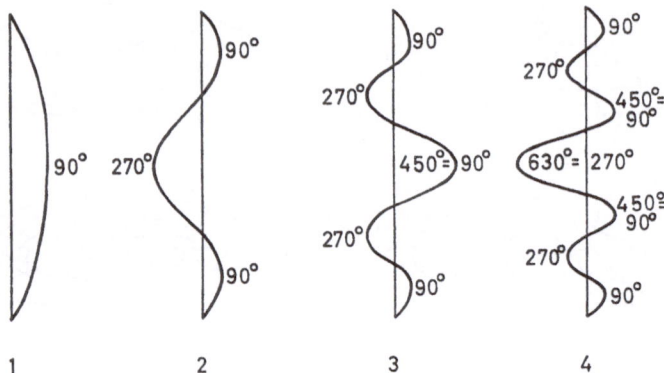

Fig. 2. The first four circularly symmetrical modes of vibration in a membrane. The approximate phase lags (between driving force and membrane displacement) are indicated at resonance for the different parts of the membrane. (*1–3* are identical to a–c on Fig. 1)

resonance is reached at a frequency (f_{02}), which is about 2.3 times f_{01}. Fig. 1b shows the membrane at the second resonance, and Fig. 2, *2*, illustrates the phase lags found at the resonance of the second mode of vibration. The movement of the membrane has now been further delayed relative to the driving force, and at a somewhat higher frequency a new nodal line will be formed, and so on (Fig. 1c shows the membrane at the resonance of the third mode of vibration, and Fig. 2 (*3* and *4*) the phase lags for the different parts of the membrane at resonance of the third and fourth mode).

In a slightly damped membrane (i.e. with a small amount of friction) the amplitude of a point on the membrane will vary a great deal over the entire frequency range. Generally, the amplitude will be a maximum at the resonance frequencies, and a minimum at a certain frequency between the resonance frequencies; but obviously this is not true for the displacement of the parts of the membrane which are at or near to a nodal circle at resonance. The variation in the overall amplitude and phase of the membrane will be fairly smooth. The amplitude and phase of individual points on the membrane, however, depend upon their position and upon the position of the nodal circles at that frequency. Thus, their variation with frequency may be rather complex. Since the receptor cells attach to small areas on the membrane, we shall consider this problem in more detail below.

It can be shown (see Morse, 1948) that if the membrane is circular, homogeneous, and acted upon by an evenly distributed force (sound wave), only these circularly symmetrical modes of vibration are to be expected. If one or more of these conditions are not fulfilled, the vibrations may be considerably more complex (Fig. 1d–f).

Methods

1. Laser Holography

Time average holograms were produced at the Division of Production Engineering, Royal Institute of Technology, S-10044 Stockholm. In the following a short description of the technique and theory of holography will be given, in order to

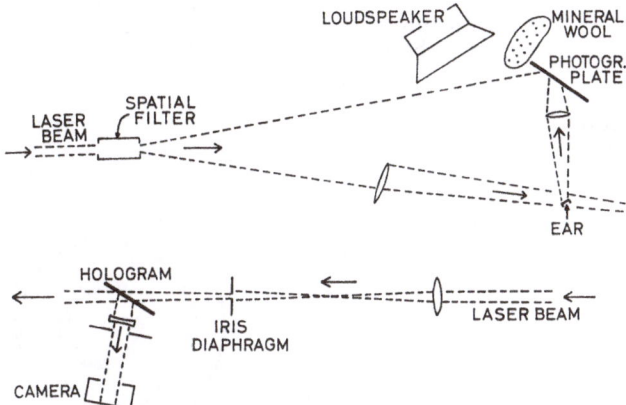

Fig. 3. The recording of the hologram (above) and reconstruction of the holographic picture (below). Explanation in the text

provide a background for an evaluation of the results. A description of the experimental technique has been published separately (Abramson, Andersson and Bjelkhagen, 1970).

The experimental set-up is illustrated in Fig. 3. The upper part of the figure shows the recording of the hologram. A laser beam, produced by a 60 mW Helium-Neon laser (Spectra Physics 125), was passed through a shutter (not shown on the figure) for 0.1–0.5 second and into a spatial filter (Spectra Physics 332). From there the light spread out, some of it reaching the photographic plate (Scientia 10E 70) directly, and some passing through a lens (+ 15 cm) and reaching the isolated locust ear. Some of the reflected light from the ear then passed through another lens (Meopta Belar, 75 mm) and on to the photographic plate.

Thus, the photographic plate received light both directly and by reflection from the ear. The light from these two sources interfered to produce a hologram. A loudspeaker was mounted near to the ear, but separated from the other parts (in order to avoid the transmission of vibrations). The distance from the ear to the photographic plate was 15 cm, i.e. more than half a wavelength of the lowest sound frequency used (1.8 kHz). An open cage of mineral wool, placed behind the ear and not shown on the figure, reduced spurious reflections of sound waves.

The lower part of Fig. 3 shows the reconstruction of the holographic picture (image). After development the photographic plate (hologram) was placed in a laser beam, which had passed a lens and an iris diaphragm. Under these conditions the hologram acts as a diffraction grating, and the wavefront reflected from the object will be reconstructed (see Pennington, 1968; Gabor and Stroke, 1969). Two pictures will be formed: a virtual image and a real image. The real image can be photographed by means of a normal camera (without a lens). In the present experiments the light from the hologram passed through a lens (–200 cm) and another iris diaphragm before reaching the camera.

If nothing moved during the recording of the hologram, the picture will look like a normal photograph (though often of bad quality). If a part of the object moved during the recording, the diffraction pattern will change. This may lead to blurred pictures, if the movement is irregular. Regular, oscillatory movements,

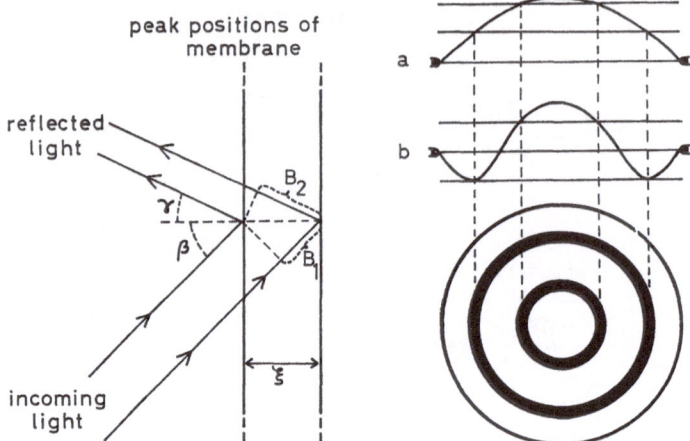

Fig. 4. *Left:* the geometry of light reflection during the recording of the holograms (see the text). *Right:* two different vibrations (*a* fundamental mode, *b* second normal mode), causing identical patterns of light and dark lines in the holographic picture (below)

however, may lead to the occurrence of fringes. Such fringes (light and dark lines) are loci of equal amplitude of vibration, corresponding to the probability density function of the motion (see Powell and Stetson, 1965; Stetson, 1970). Light reflected from the two peak positions of the oscillating membrane will interact; the intervals of peak-to-peak displacement corresponding to the occurrence of fringes depend on the wave length of laser light and on the angles of incidence and reflection of the light upon the tympanal membrane (see Fig. 4). If the peak-to-peak amplitude is ξ, and the angles of incidence and reflection are β and γ, one has

$$B_1 + B_2 = \xi \cos \beta + \xi \cos \gamma. \qquad (1)$$

When $(B_1 + B_2)$ is equal to (a multiple of) the wavelength of the light (λ_1), the two reflected beams are in phase. The corresponding peak-to-peak vibration amplitude (ξ') becomes

$$\xi' = n \cdot \frac{\lambda_1}{\cos \beta + \cos \gamma} \qquad (2)$$

where $n = 0, 1, 2, \ldots$.

In the present case $\beta = 45°$ and $\gamma = 30°$. The wavelength of the laser light was 0.6328 μm. Thus, normally the distance between two dark bands or between two light bands on the holograms would correspond to a difference in peak-to-peak amplitude of 0.40 μm. When the fundamental vibration is observed, one can determine the displacement of the centre simply by counting the number of rings (Fig. 4, *a*). This is, however, not the case when higher, circularly symmetric modes of vibration are observed. In such modes of vibration the concentric "bubbles" of the membrane may each give rise to one or more rings (Fig. 4, *b*). Although the nodal lines may have special contours, this is not always the case. The position of the nodal lines in the locust ear could not be determined directly from the holograms, but had to be determined with the capacitance electrode measurements.

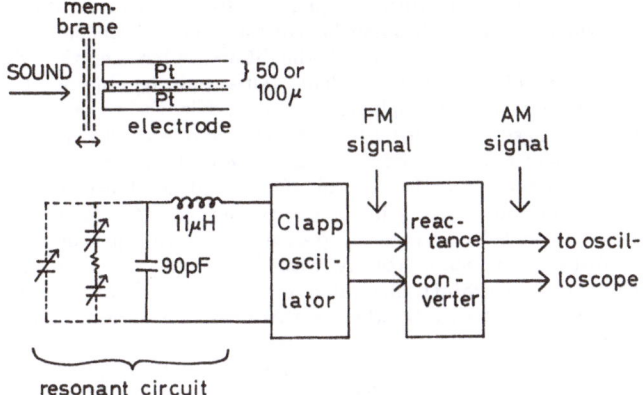

Fig. 5. The capacitance electrode (above), and the principle of the capacitance measurement (below). Explanation in the text

2. The Capacitance Electrode

The vibration of small areas of the tympanal membrane was measured by means of a capacitance electrode. It consisted of two platinum wires (50 or 100 μm in diameter) glued closely together with araldite for a distance of 1 cm. The wires were cut (at right angles to their longitudinal axis) and laquered with Insl-X. The free ends of the wires were connected to a capacitance transducer system (DISA 51 D 17). When the tip of this electrode was placed near to a moving membrane (Fig. 5), the (periodic) change in capacitance between the membrane and the two wires changed the resonance frequency of a resonant circuit coupled to a Clapp oscillator (DISA 51 E 02). The frequency-modulated signal was demodulated in a reactance converter (DISA 51 E 01), and the AM-signal displayed on the oscilloscope. This system makes it possible to measure periodic changes in capacitance down to 10^{-3} pF (with a small and known phase-shift) in the frequency range 0–200 kHz.

The absolute amplitude of vibration could not be estimated with this method, since the magnitude of the signal depends both on the effective size of the corresponding area of membrane, on the distance to the electrode tip, and on the electrical properties of the membrane. [The effect is probably a combination of two processes: the moving membrane changes the dielectric properties in the field between the electrode tips, but at the same time the cell layer on the membrane is conducting, so that the system can also be regarded as composed of two variable capacitances connected by an (unknown) resistor, see Fig. 5].

In order to allow the phase relationships to be measured, the isolated ear and a microphone (Brüel & Kjär 4131) were placed at the same distance from the loudspeakers (about 1.3 m). The sound signal was amplified and displayed on the oscilloscope together with the signal from the capacitance electrode system. The sensitivity of the system was sufficient to give signals with a signal/noise ratio between 1 and 10, when the 100 μm electrode was used. The signals obtained by means of the 50 μm electrode were often too small compared with the noise level. Obviously, a more sensitive system would be needed for studies on most other insect ears (a 10 times more sensitive oscillator is now under construction at DISA).

No filtering was performed, because the phase shifts in selective filters are much too drastic to allow accurate determinations of the phase relationships.

The phase difference of interest here is the difference between the driving force and the displacement of the membrane. This value could be obtained by allowing for the phase shifts in the microphone system and in the reactance converter. Furthermore, the phase difference between the sound pressure wave (measured with the microphone) and the force (acting to move the membrane) had to be calculated. In the following article (paper III) it will be shown how this figure was derived. The total uncertainty in the determination of the force-displacement phase difference depended on frequency (because the errors produced by a small difference in distance increased linearly with frequency). At "low" frequencies (1–2 kHz) the total uncertainty amounted to about $\pm 10°$; at high frequencies (15 kHz) it was about $\pm 20°$. Nevertheless, this degree of accuracy was sufficient for the present purpose.

It is relatively easy to use the capacitance electrode on plane parts of the tympanum. The attachment areas of the receptor cells, however, are so irregularly shaped that recordings were difficult. In order to interpret the results it is necessary to have an idea of the kinds of amplitude and phase relationships which can be expected from small areas on various parts of the membrane (see below). Since this method does not tell us the absolute amplitudes, the results are most easily interpreted when compared with the laser holograms (the holograms, on the other hand, do not tell us anything about the phase relationships).

The Expected Resonances

1. Properties of the Tympanal Membrane

Anatomy. The anatomy of the tympanal organ has been described in detail by several authors (Gray, 1960, gives a list of references). In the following description only the properties of relevance to the problem of membrane vibration will be considered. The tympanal membrane is bean-shaped (Fig. 6), and the dorsal end is somewhat wider than the ventral end. Typically, the membrane is 2.5 mm long and 1.5 mm at its widest. The total area is about 3 mm². Most of the tympanal membrane consists of cuticle (2–3 μm thick), which is covered on the inside by a layer of cells (1–2 μm thick) and by the wall of an air sac. This *thin part* of the membrane has an area of about 2.4 mm². In the ventral anterior corner, however, the membrane is about 8–10 μm thick. The area of this *thick part* of the tympanum is about 0.5–0.6 mm² (Fig. 6).

Four specialized regions with much thicker cuticle are situated between the middle of the tympanum and the anterior edge. They are the elevated process, the styliform body, the folded body, and the pyriform vesicle. The four anatomical groups of receptor cells (a, b, c and d) are attached to these bodies (Fig. 6). The areas of attachment of the a, b, and c-cells are situated at the junction of the thin and thick parts of the membrane and fairly close to each other. The d-cells, however, are attached to the pyriform vesicle 200–300 μm from the rest of Müller's organ. The thin and thick parts of the tympanum are separated by the

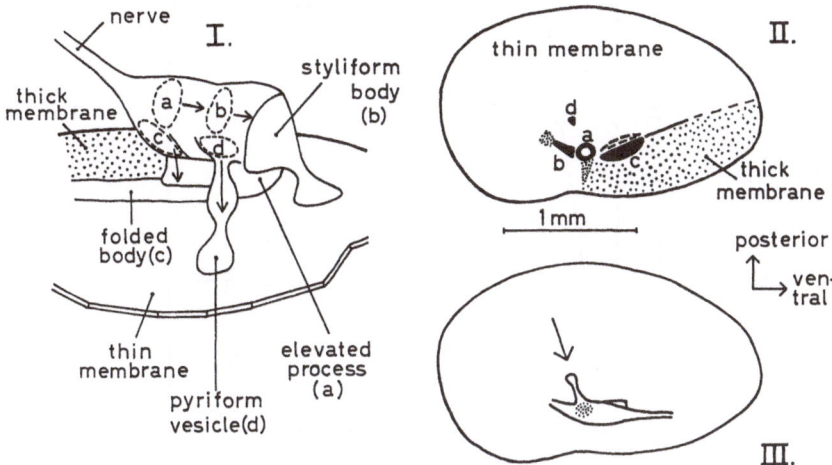

Fig. 6. The anatomy of the tympanal organ. *I* Müller's organ (right ear). The position of the receptor cells (a–d) and the attachment parts of the membrane are shown. Arrows indicate the direction of the dendrites. *II* The left ear seen from outside the animal. The dark areas indicate the areas of attachment of the receptor cells. *III* The right ear seen from the inside. The arrow indicates the visual angle of *I*

folded body and by a rod-shaped thickening in continuation of the folded body.

Mass. The mass of the tympanal membrane was determined with a microbalance (Cahn Gram Electrobalance). The membrane was cut out with fine scissors and kept in a moist chamber until the microbalance had been calibrated. The average weight of 13 tympanal membranes was 30 μg (s.d. = 1.0 μg). In three other cases, however, weights around 50 μg were found. Müller's organ in these preparations appeared swollen. The average weight of the thin part of the membrane was 7.7 μg (10 determinations, s.d. = 1.7 μg).

These values may be compared with the calculated weights: If one assumes a specific weight of 1.2 for the cuticular part of the membrane (Jensen and Weis-Fogh, 1962) and about 1 for the cell layer, the weight of the thin membrane should be about 10.8 μg. The weight of the entire membrane is more difficult to estimate, because of the irregular shape of Müller's organ, but 40–45 μg would be an approximate estimate. This difference between the determined and estimated weights is probably due to drying of the preparation during the operation and weighting procedure: Typically, a tympanum with an initial weight of 30 μg lost about 1–2 μg of water during the first three minutes. Although the operation and determination of the weight were carried out quickly,

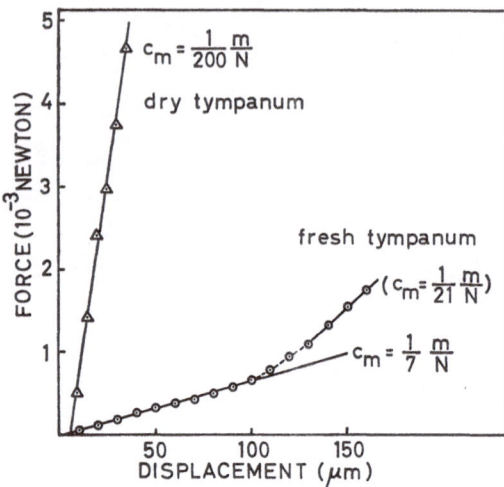

Fig. 7. A measurement of the membrane compliance (c_m). The lower curve is from a fresh tympanum, and the upper curve from the same preparation 18 hours later. Note that the compliance of the fresh membrane is almost constant for the first 100 μm displacement

the values obtained are probably somewhat too low. In the calculations performed below values intermediate between the estimated and determined weights (36 μg and 9.2 μg, respectively) will be used instead of the determined weights. Without this correction the calculated frequencies would have been about 10% higher.

Tension. The elasticity of the tympanal membrane was determined by displacing it in the perpendicular plane with the end of a metal wire. The wire had a diameter of 200 μm, and it was attached to a capacitive transducer (DISA 51 D 17) fitted with a specially made transducer plate. The capacitance transducer was mounted on a screw gauge, which could be moved in 5 or 10 μm steps. The determinations were carried out quickly in order to avoid errors due to "plastic flow" in the membrane.

The mechanical compliance (c_m) was about 0.14 m/N (12 determinations, s.d. = 0.023). In Fig. 7 the result of a typical experiment is shown. The relationship between displacement and force is approximately linear for displacements up to 50–100 μm i.e. in this range Hooke's law is obeyed, and the compliance is constant. Above this range the force increases rapidly, i.e. the compliance decreases.

The restoring force of a surface may be due to tension and/or stiffness of the surface material. In the present case (fresh tympanum, see below), the tension seems to be the dominating factor, since the compliance for the thin part of the membrane did not differ significantly from that of the attachment areas of the receptor cells. Also, when slits were cut in the tympanum, the holes adopted the shape of a convex lens, and free tympana were quite lose. A surface whose compliance is mainly determined by its tension, is physically speaking a membrane (in contrast to a plate, in which the stiffnes is an important factor).

A considerable decrease in compliance was observed if the membrane was allowed to dry. After some hours in a dry environment the compliance may be about $1/2-1/30$ of the value for a fresh membrane (Fig. 7). This decrease in compliance is largely due to a greatly increased stiffness (cf. Herzog, 1926; Jensen and Weis-Fogh, 1962). Since a decrease in compliance will tend to shift the resonance frequencies to higher values, it is very important to keep the preparation under humid conditions during experiments.

2. The Expected Resonance Frequencies

The fundamental resonance frequency (f_{01}) in vacuum for a homogeneous membrane, fastened along a boundary circle, is given by (see Morse, 1948)

$$f_{01} = 0.383 \cdot \frac{1}{a} \sqrt{\frac{T}{\sigma}} \qquad (3)$$

where a = radius (m),
   T = tension per unit length (N/m),
   σ = mass per unit area (kg/m² = N s²/m³).

In the present case the use of formula (3) is questionable, since the membrane is neither in vacuum, circular, nor homogeneous.

The tympanal membrane is far from homogeneous. Although the thin, homogeneous membrane constitutes more than 80% of the area, about two-thirds of the total mass is concentrated in Müller's organ. Furthermore, the presence of the four cuticular bodies in the tympanum are likely to affect its vibration. These bodies are stiff, and they cover most of the boundary between the thin and thick parts of the tympanum. The presence of this stiff boundary at the edge of the thin membrane may have the effect that *the thin membrane vibrates independently of the entire tympanal membrane*. In the calculations it is assumed that this is actually the case, and that the thin part of the membrane has a series of resonance frequencies of its own. The laser holograms show that this assumption is correct (see below).

From the area of the thin and entire membrane the probable effective radius can be estimated to 0.9 and 1.0 mm, respectively. According to Lord Rayleigh (1926) the fundamental frequency of a membrane is determined by its area rather than by its shape, but the estimate of the effective radius (a) is still uncertain. Since the vibrations take place in air, the radiation impedance of the membrane must be considered. Thus, the mass per unit area (σ) is estimated from the mass of the membrane plus the radiation mass (see the Appendix). The membrane tension (T) can be calculated from the compliance (c_m). If the membrane is pulled evenly around its edge with a tension of T Newton per meter of edge, it can be shown (see Morse, 1948) that for a circular membrane

$$T \simeq \frac{1}{8 \cdot \pi \cdot c_m} \tag{4}$$

In the present case T becomes 0.28 N/m.

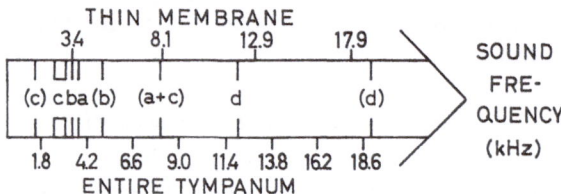

Fig. 8. The expected resonance frequencies for the thin membrane (above) and for the entire tympanum (below), compared with the preferred frequencies of the four groups (a–d) of receptor cells (centre). Letters in brackets indicate frequencies not preferred by all cells in a group

From these figures and Eq. (3) the fundamental resonance frequency (f_{01}) should be expected at about 1.8 kHz for the entire tympanum and at about 3.4 kHz for the thin part of the tympanum. In vacuum the resonance frequencies of the second, third, and fourth circularly symmetric modes are expected at 2.296, 3.599, and 4.903 times f_{01}. In air the variation of the radiation mass has to be taken into account (see the Appendix). The expected resonance frequencies of ideal membranes with size, mass, and compliance equal to that of the entire and thin tympanal membranes are listed on Fig. 8.

In Fig. 8 a few of the first resonances are indicated. There are, of course, theoretically an unlimited number of resonances, but for the higher modes the distance between the nodal lines becomes very small. The receptor cells do not attach to points on the membrane, but to stiff, cuticular bodies which occupy certain areas. Therefore, the modes of vibration are unlikely to excite the receptor cells, if the distance between the nodes is much smaller than the diameter of the attachment areas.

The resonance frequencies may be compared with the frequencies of maximum sensitivity in the receptor cells (Fig. 8). It is seen that most of the observed best frequencies are fairly close to one of the expected resonance frequencies. On the other hand, the number of expected resonances is so large in the frequency range 1–20 kHz that a wide variety of explanations may be given for several of the observed best frequencies. The problem also arises how the receptor cells can be so selective: each group of receptor cells responds to only two or three bands of frequencies. Obviously, direct observations of the vibrations are necessary.

3. Expected Amplitude and Phase of a Point on the Membrane

In most studies of membrane vibration, special interest is paid to the average (overall) phase and amplitude of the membrane, since it is the average that determines the output from loudspeakers or microphones. In the locust ear, however, the receptor cells attach to cuticular bodies which occupy certain areas of the tympanum (Fig. 6). In the following we shall first calculate the amplitude and phase of single points on a membrane in order to obtain an idea of the kind of behaviour to be expected for the areas of attachment. Thereafter, the interaction between two vibrations will be considered.

We shall consider a homogeneous, circular membrane with physical properties (size, mass, tension, and friction) almost identical to those of the thin membrane in the locust ear. Since the friction is determined mainly by the membrane material (see the appendix), the equation of motion becomes (see Skudrzyk, 1954, p. 331)

$$T\left(\frac{\partial^2 \eta}{\partial r^2} + \frac{1}{r}\frac{\partial \eta}{\partial r}\right) + \frac{F}{\pi a^2} = \sigma \frac{\partial^2 \eta}{\partial t^2} + \frac{R}{\pi a^2}\frac{\partial \eta}{\partial t}, \tag{5}$$

where η = amplitude of displacement (m),
$\quad T$ = membrane tension (N/m),
$\quad r$ = distance from centre (m),
$\quad a$ = radius of membrane (m),
$\quad F$ = the force acting to move the membrane (N),
$\quad \pi = 3.14159\ldots$,
$\quad \sigma$ = mass per unit area (kg/m²),
$\quad t$ = time (s),
$\quad R$ = resistive component of membrane impedance (Ns/m).

It can be shown (Crandall, 1927) that the solution of Eq. (5) giving the displacement of a point on the membrane is

$$\eta = \frac{F \cdot (J_0(k_1 r) - J_0(k_1 a))}{T \cdot \pi \cdot a^2 k_1^2 \cdot J_0(k_1 a)} \, e^{j\omega t}, \tag{6}$$

where $k_1^2 = \omega^2 \cdot \dfrac{\sigma}{T} \cdot \left(1 - j\,\dfrac{R}{\pi a^2 \omega \sigma}\right)$,
$\quad j = \sqrt{-1}$,
$\quad J_0$ = Bessel function of complex argument of order zero for cylindrical coordinates,
$\quad \omega$ = angular frequency ($= 2\pi f$),
$\quad e = 2.71828\ldots$, base of natural logarithms.

From Eq. (6) the amplitude of vibration and the phase relations can be found for different distances to the centre. The variation of force with frequency (see paper III) was also taken into account in the calculation of the amplitudes. (If the membrane were a part of a pressure receiver, the curves would indicate the relative velocity of vibration at different frequencies; in this case the corresponding amplitudes can be found by dividing with ω; i.e. with increasing frequency the amplitudes will decrease 6 dB per octave.) Since Eq. (6) is fairly complicated, the solutions were calculated by means of a digital computer.

The behaviour of ideal membranes was described in the introduction and illustrated on Figs. 1 and 2. The vibrations of membranes governed by Eqs. (5) and (6) differ somewhat from those of ideal membranes. The most important difference is that there are no real nodal circles on the tympanum: all parts of the tympanum (except the edge) are moving, but the amplitude of vibration is a minimum at the parts corresponding to the nodal circles in the ideal membrane. The membrane on both sides of these "relative nodal circles" is vibrating about 180° out-of-phase, but the change of phase is continuous. In contrast, in the ideal membrane the change of phase across the nodal circles is discontinuous. This difference between the ideal membrane and the tympanum is interesting, but it does not change the situation of the receptor cells very much. The receptor cells are attaching to cuticular bodies occupying certain areas on the membrane and not to infinitely small points. The presence of relative instead of real nodal lines will not cause a great difference in the vibration of the areas of attachment. Therefore, for the present purpose the nodal circles can still be treated as a reality.

Another difference from the ideal behaviour is found in the definition of resonance. In the ideal system the resonance frequencies may be defined as the frequencies at which the amplitude of vibration of the centre is a maximum; but it may also be defined as the frequencies at which the phase lag of the centre is $90° + n\,180°$ (where $n = 0, 1, 2, \ldots$). In the vibration determined by Eqs. (5) and (6) these two definitions give different resonance frequencies. When the frequency is increased at the second mode of vibration the amplitude is a maximum about 100 Hz before the phase lag reaches 270°. Here again, the deviation from the ideal behaviour has no practical significance.

In Fig. 9 the expected amplitude and phase lag at different frequencies are shown for the centre of a membrane with thin-membrane properties. The damping factor (R) was approximately that estimated from the holograms (see below). The variation in radiation mass with the mode of vibration (see Appendix) was, however, not taken into account. Therefore, the resonance frequencies are not quite identical with those indicated on Fig. 8. These curves show that the phase lag of the centre is about 90° at resonance of the fundamental mode, and that the phase lag increases about 180° for each mode (cf. Fig. 2). It is surprising that the amplitude at 5 to 6 kHz is only 3–4 times smaller than the amplitude at the first two resonances. The computed amplitude and phase curves for the centre of the entire membrane are very similar to those shown on Fig. 9. The entire membrane system is, however, about 3 times more damped than the thin membrane. The variation in vibration amplitude is therefore not so large. The amplitude of the first resonance (at 1.8 kHz) is here only 2 times larger than that at 3 kHz (cf. Fig. 15).

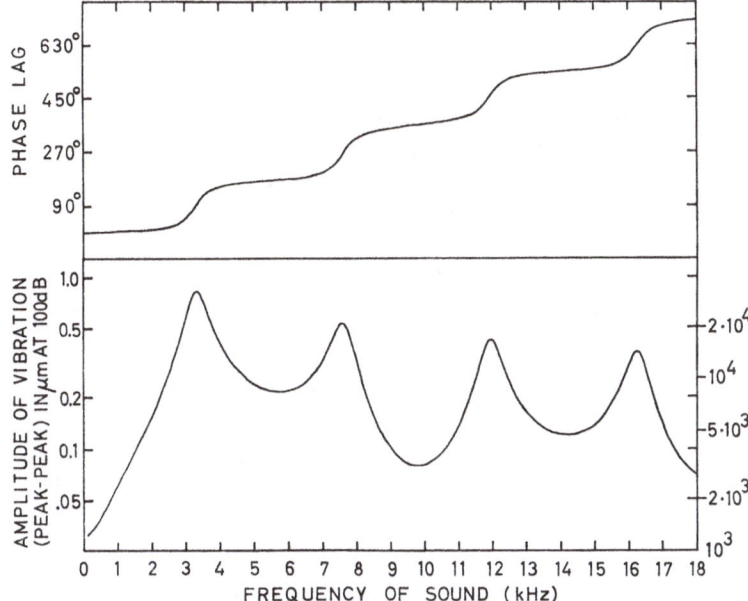

Fig. 9. Expected amplitude of vibration (below) and phase lag (above) for the centre of the thin membrane. The peak-peak amplitudes indicated in μm to the left are valid for the isolated ear at a sound level of 100 dB. The scale to the right gives the amplitudes as $\eta \, \omega/F$. The resonance frequencies are not quite correct, since the effective mass was kept constant (see the text)

The phase lags computed for points elsewhere on the membrane show an increase similar to that of Fig. 9. A decrease of about 180° in the phase lag is, however, noted each time a nodal circle passes the point in question. The variation in amplitude with frequency may be very different for different points on the membrane. In order to illustrate this difference the expected amplitudes were computed for different points on the membrane at the second mode of vibration (Fig. 10). The numbers indicate the fractional distance to the centre. It is seen that the vibration of the centre is a maximum at the resonance frequency (cf. Fig. 9). In contrast, the vibration of points with a fractional distance of about 0.44 from the centre is a minimum at the resonance frequency, i.e. these points are on the nodal circle at resonance. Also, it should be noted that parts of the membrane near to a nodal circle may show a maximum vibration amplitude at a frequency somewhat below or above the resonance frequency. This means that the frequency of maximum sensitivity in receptor cells attached to such parts does not necessarily have to be identical with the resonance frequency.

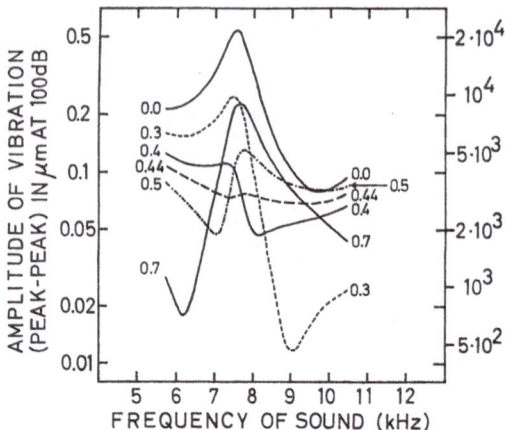

Fig. 10. The expected amplitudes at the second mode of vibration of the thin membrane. Numbers on the curves indicate the fractional distance to the centre (r/a). Note the position of the nodal circle at different frequencies

So far, we have only considered the behaviour of points on an ideal membrane with one set of vibrations. In such a system the modes follow one after the other, a new mode of vibration gradually being "born", when a new nodal circle is formed. When such a membrane is moved by a pure tone there is, of course, no possibility for any interactions between the modes, since they never exist simultaneously. It should be borne in mind, however, that the tympanum is not an ideal membrane; the attachment areas are not points on a homogeneous membrane; and there are at least two sets of vibrations.

We shall now try to obtain an idea about the expected amplitude and phase of a point on the membrane, which participates in two different vibrations (1 and 2) at the same time. If the point took part in vibration (1) only, it would be displaced with an amplitude D_1 and with a phase lag (relative to the driving force) ε_1, and similarly vibration (2) would give D_2 and ε_2. The magnitude of these displacements and their phase lags can be drawn by means of two vectors (Fig. 11). We now assume that the effect of the two vibrations can be added simply by adding the two vectors. By means of simple trigonometry one finds that

$$D_3 = \left(D_1^2 + D_2^2 + 2D_1 D_2 \cos(\varepsilon_2 - \varepsilon_1)\right)^{1/2}, \tag{7}$$

$$\mathrm{tg}(\varepsilon_3 - \varepsilon_1) = \frac{D_2 \sin(\varepsilon_2 - \varepsilon_1)}{D_1 + D_2 \cos(\varepsilon_2 - \varepsilon_1)} \tag{8}$$

where D_3 and ε_3 are the amplitude and phase lag of the resultant vibration of the point in question.

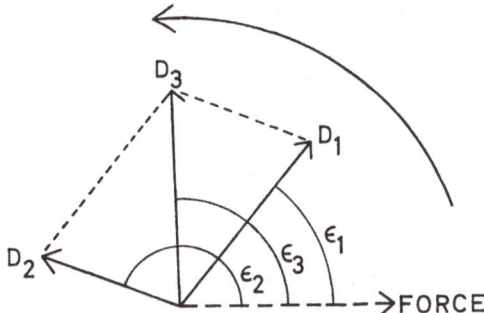

Fig. 11. The addition of two simultaneous vibrations (1 and 2). The displacement (D) and phase lag (ε) are drawn as vectors. D_3 and ε_3 indicate the resultant vibration

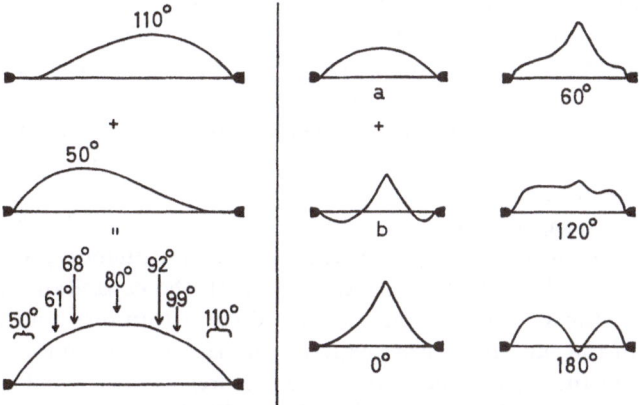

Fig. 12. Simple examples of the addition of two vibrations by means of Eqs. (7) and (8). *Left:* Different phase lags of the two components result in a gradual change in phase over the membrane. *Right:* The addition of the vibrations a and b may result in very different vibrations depending on the phase lags of the components (numbers on the four resultant vibrations indicate the difference in phase between the centers of a and b)

In Fig. 12 these equations have been used to calculate the amplitude and phase of some resultant vibrations. At left, two vibrations with different phase lags are added. The phase lags of the resultant vibration are changing gradually over the membrane. At right, the vibrations a and b have been added. In this example the difference in phase lag between the two vibrations has been varied. It is seen that very different resultant vibrations are obtained at various differences in phase between the component vibrations.

Naturally, these examples are over-simplifications. In the tympanal membrane both amplitude and phase of the component vibrations vary with frequency. At most frequencies the difference in phase between the component vibrations remains relatively constant. Around the resonance frequencies, however, both amplitude and phase change rapidly. Note that the addition of the two vibrations may lead to the formation of an "artificial nodal line" (Fig. 12, right).

The Observed Vibrations

1. Properties of the Tympanal Resonances

In the preceding paragraphs the expected vibrations have been calculated on the assumption that the tympanal membrane behaves as an ideal membrane. It is obvious that the entire tympanum is very far from being homogeneous, and that the thin part of the membrane (although fairly homogeneous) is not circular. The most surprising feature of the vibrations, as observed by means of laser holography and the capacitance electrode, is how well they follow the predicted behaviour. There are, of course, several minor deviations between the predicted and the observed vibrations, but the overall impression is that of a general agreement.

The following description of the vibration patterns in the isolated ear is based on a total of 68 holograms (of 7 different preparations) and a total of about 1500 measurements with the capacitance electrode (53 preparations). All experiments were done within one hour after the removal of the ear. In the laser experiments the preparation was covered with humid filter paper between the exposures.

The best correlation with theory is found for the resonance frequencies; in fresh preparations most resonance frequencies deviate less than 10 per cent from the expected values. It is, however, difficult to give exact measurements for the resonance frequencies, since there is some variation between the preparations. Also, somewhat higher values were found at the end of the experiments, if the preparation had been allowed to dry. The effect of drying was relatively moderate (about 10–15 per cent increase in the resonance frequencies during the first hour), but obviously the drastic increase in the stiffness illustrated in Fig. 7 is a warning against experiments of longer duration. The variation between the preparations and the effect of drying seems to be sufficient to explain the variation observed in the recordings from single receptor cells (paper I).

The most surprising deviation from theory is that the spatial position of the centre of vibration is not constant. In an ideal membrane the centre of vibration should be located at the geometrical centre of the

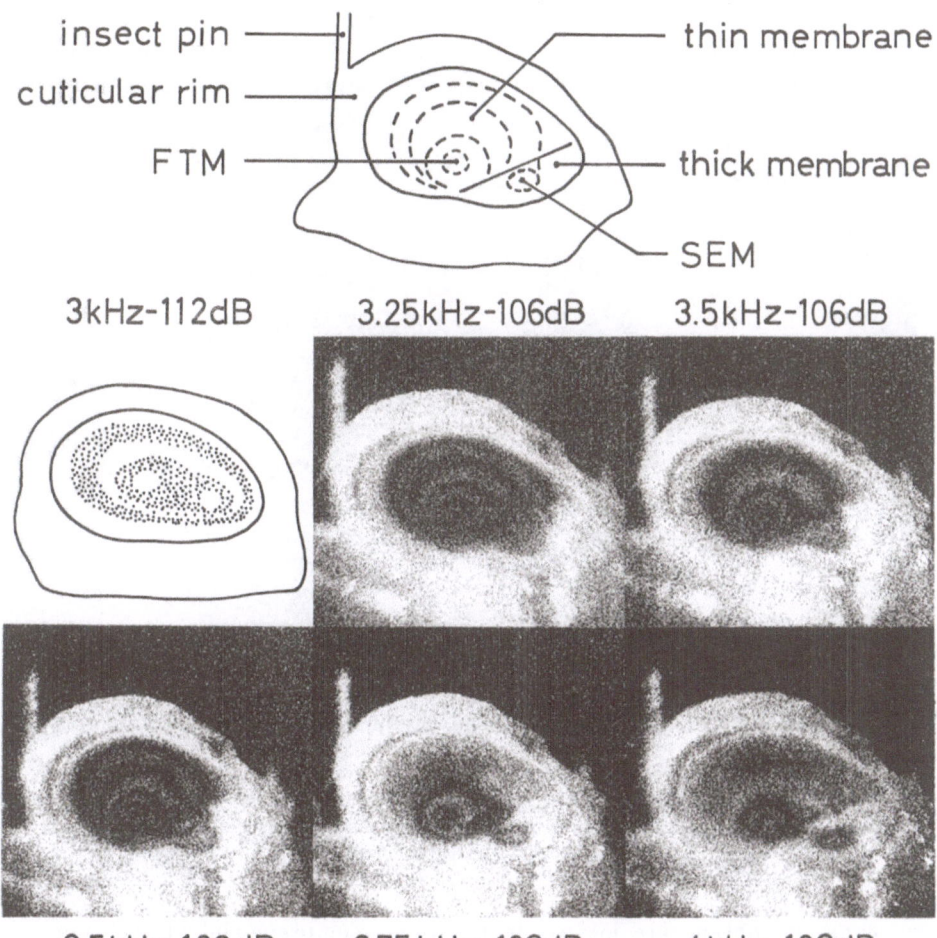

Fig. 13. Holographic pictures of the vibrations in the frequency range 3–4 kHz. *Above:* The orientation of the isolated ear, and the approximate positions of the centers of the fundamental mode of the thin membrane (FTM) and of the second mode of the entire membrane (SEM). *Below:* The vibration patterns; the dark and light lines on the pictures are loci of equal amplitudes of vibration (see Fig. 4). Note the concentration of the FTM vibration, when the frequency is varied from 3 to 3.75 kHz. (The 3 kHz vibration pattern has been drawn from a holographic picture of bad technical quality. This picture was from another preparation)

membrane. In the tympanal membrane, however, the centers of vibration are not at the geometrical centre, and their positions differ for different modes of vibration. The centers of all the modes of vibration of the

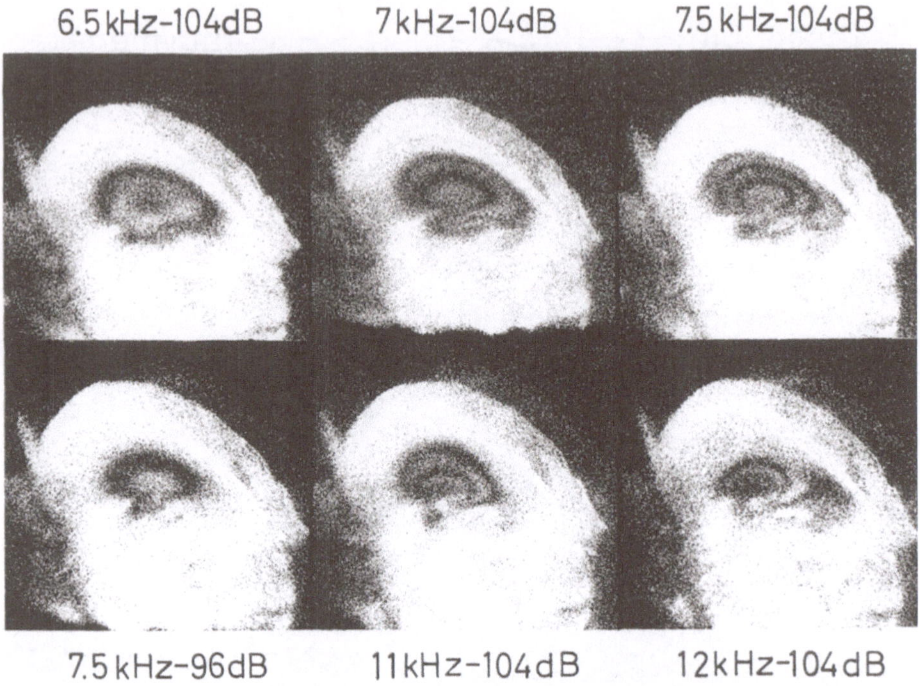

6.5 kHz–104dB   7 kHz–104dB   7.5 kHz–104dB

7.5 kHz–96dB   11kHz–104dB   12kHz–104dB

Fig. 14. Holographic pictures of the second and third mode of vibration of the thin-membrane system. The orientation of the isolated ear is similar to Fig. 13. The three upper pictures show the increasing amplitude of vibration as the second resonance frequency (8 kHz) is approached. The pictures at 11 and 12 kHz show the concentration of the area of vigorous vibration as the third resonance frequency (13 kHz) is approached

entire membrane are located in the thick-membrane-end of the tympanum; but the centre of the first mode is at the thin membrane (Fig. 15), whereas the centers of higher modes are in the thick membrane.

The centers of vibration of the *thin*-membrane system are located nearer to the geometrical centre of the tympanum (Figs. 13 and 14). Here again, the position of the centre is not constant; at some frequencies the centre of vibration is at the geometrical centre of the thin membrane (Fig. 14); at other frequencies it is at the area of attachment of the a- and b-cells, i.e. near the edge of the thin membrane (Fig. 13). This may seem surprising, but because of the peculiar shape of the thin membrane (Fig. 6) the area of attachment of the a- and b-cells is in fact both at the edge of the thin membrane and near to its centre.

This difference in the spatial position of the two sets of vibrations is also evident from the measurements with the capacitance electrode.

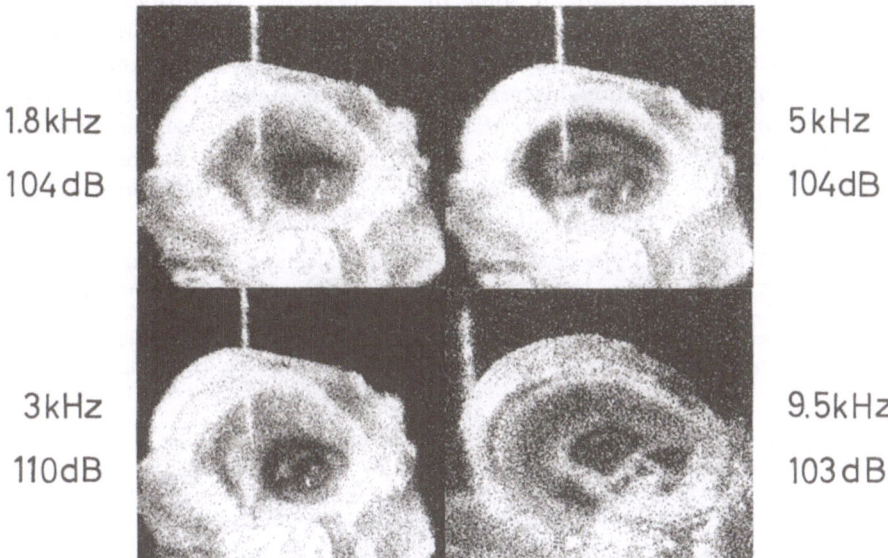

1.8kHz
104dB

5kHz
104dB

3kHz
110dB

9.5kHz
103dB

Fig. 15. Holographic pictures of vibration patterns in the isolated ear. *Left:* The fundamental mode of the entire-membrane system. Note that the centre of vibration is in the thick-membrane corner of the tympanum. *Right:* Two unexpected patterns of vibration (see the discussion). In three of the pictures the insect pin is also seen on a part of the membrane. In fact, the pin was behind the membrane, and reflected light was transmitted through the membrane

When the electrode is placed at the thick-membrane-end of the membrane, the phase lag and the variation in amplitude are close to those expected for the entire-membrane system. In the opposite end of the membrane, however, the values are close to those expected for the thin-membrane system.

The behaviour of the membrane between the two centers of vibration is determined by both vibrations (see above). In the larger part of this "zone of interaction" the final vibration is determined mainly by the vibrations of the thin-membrane system. This is not surprising, since the damping of the thin membrane is about three times less than that of the entire membrane (see below). Thus, above 3 kHz the amplitude of vibration of the thin-membrane system is normally larger, and consequently this system will tend to "dominate" the entire-membrane system. Some of the interactions will be described in detail below, since the attachment area of some receptor cells is in the "zone of interaction" between two vibrations.

A surprising feature of some of the tympanal vibrations is their spatial concentration. It was mentioned above that the vibrations of the entire membrane have their centre in the thick-membrane-end of the tympanum. Although these vibrations are derived from the entire tympanum, vigorous movement may be restricted to a very small area (SEM on Fig. 13). This is also true for some of the thin-membrane vibrations, e.g. the fundamental mode at 3.75–4 kHz (FTM on Fig. 13) and the third mode at 12 kHz (Fig. 14).

The Vibration at Different Frequencies. On the holograms at 1.8, 2, and 2.5 kHz the fundamental mode of the entire membrane (FEM) covers a fairly large part of the thick-membrane-end of the tympanum. At 3 kHz the FEM may still be the only vibration seen (Fig. 15), but in some preparations and at high intensities the fundamental mode of the thin membrane (FTM) can also be seen (Fig. 13). The capacitance electrode experiments show how these vibrations interact (see below: c-cells).

In the frequency range 3.25 to 4.5 kHz the vibrations are dominated by the FTM. The second mode of the entire membrane (SEM) is now restricted to a small area on the thick membrane. The FTM-vibration may be fairly uniform over the membrane (3.25 kHz on Fig. 13), but generally it is spatially concentrated (Fig. 13). Therefore, it is not surprising that the FTM seldomly causes a clear maximum of vibration in the capacitance measurements. These vibrations are discussed below (a- and b-cells).

At 5 kHz the holographic picture may be very similar to that at 4 kHz (Fig. 13), but in one preparation the picture of Fig. 15 was obtained. The nature of this vibration is discussed below.

At 6.5 to 8.5 kHz the vibration is dominated by the second mode of the thin membrane (Fig. 14). This vibration is well suited for detailed measurements with the capacitance electrode, since its interactions with the entire-membrane vibrations seem to be minimal. The position of the nodal circle at different frequencies could be determined by placing the electrode at different positions on the membrane and gradually increasing the sound frequency. If the distance between the electrode and the centre was more than about half the radius, a sudden decrease in phase could be observed at a certain frequency below the resonance frequency. At the same time the amplitude of vibration was a minimum. At a distance of about half the radius this drop in phase and amplitude occured near to the resonance frequency (see Fig. 17). Near the center both the phase lag and the amplitude increased without interruption. These experiments confirmed the impression gained from the holograms: the position of the second mode is almost symmetric on the thin membrane. This is in contrast to the fundamental and third modes, which are very askew. The experiments also agree reasonably well with the behaviour predicted from physical theory (Fig. 10), but since the tip of the electrode was not comparable to a point, the changes of phase and amplitude were not as dramatic as expected.

At 9.5 kHz (Fig. 15) an unexpected vibration exists on a very limited area of the membrane (see the discussion). At this frequency one can also see another vibration on the holographic picture (Fig. 15); this vibration is located at the thick-membrane end of the tympanum, and probably it is the fourth mode of the entire membrane.

At 11 to 14 kHz the third mode of the thin membrane can be seen on the holograms (Fig. 14) and measured with the capacitance electrode (see below:

d-cells). At higher frequencies, however, the vibrations become too complex to be resolved by the present technique.

2. The Damping Factor and the Tuning of the Receptor Cells

In general, friction has very little influence in determining the resonance frequency, and it has therefore been neglected in the calculations of resonance frequencies performed above. The amplitude of vibration, however, is markedly influenced by the friction of the system. The friction can be divided into external and internal friction. The external friction (indicated as $R_r =$ the resistive component of the radiation impedance) is equal to the resistance to the motion which the surrounding air manifests. In the appendix it is shown that the external friction can be neglected for the isolated membrane. Thus, we shall only consider the contribution of the internal friction (indicated as $R =$ the resistive component of the membrane impedance).

The magnitude of R determines the amplitude of vibration at resonance, and it also determines the degree of tuning to the resonance frequency. The degree of tuning is often expressed by the "Q of the system". Q is the number of cycles required for the amplitude of motion to reduce to $1/e^{\pi}$ of its original value (approximately 0.043), when the driving force stops.

In a *simple oscillator* (mechanical or electrical) Q can be found from the tuning to the resonance frequency (f_0) by means of the expression

$$Q = \frac{f_0}{f_2 - f_1} \tag{9}$$

where f_1 and f_2 are the frequencies below and above f_0, at which the response of the system is 3 dB down relative to the response at the resonance frequency. The relationship between R and Q is given by

$$Q = \frac{1}{R \cdot \omega_0 \cdot c_m} = \frac{\omega_0 \cdot m}{R} \tag{10}$$

where $\omega_0 =$ angular frequency at resonance ($= 2 \cdot \pi \cdot f_0$),

$c_m =$ the compliance,

$m =$ the mass.

These equations do not apply to the higher modes of membranes, but they are almost valid for the fundamental mode of vibration, if some corrections are made: The values of c_m, m, and R cannot be used directly in the equations for simple oscillators (see Morse, 1948). Thus, the value of f_0 for the thin membrane calculated from Eq. (10) is about 4 kHz, as compared with the 3.4 kHz found by means of Eq. (3). Therefore, if the determined value of c_m is used in Eq. (10), the effective mass used should be about 1.38 times the real mass. Similarly, the value of R used in Eq. (10) will depend on the part of the membrane considered, because the membrane—unlike a simple oscillator—does not vibrate with a uniform amplitude of motion. It can be shown (see Skudrzyk, 1954, p. 333) that the value of R calculated from the amplitude of the centre of vibration is about 0.43

times the value to be used for the calculation of Q by means of Eq. (10). The real, physical R, however, has to be calculated by means of Eq. (6).

In the following, the values of Q estimated from the tuning of the receptor cells will be compared with the values expected from the amplitude of vibration observed on the holograms. The cells are primary receptor cells, and there is probably no lateral interaction between them. Therefore, the tuning of their threshold curves are likely to reflect the tuning of the vibration of their areas of attachment. The amplitude of vibration can be obtained from the holograms. The force, however, has to be estimated by means of indirect methods. In the following article (paper III) the force acting to move the tympanum in the isolated ear is calculated by comparing the sensitivity of the isolated ear with that of an operated, semi-intact ear.

The average value of (the physical) R for the thin membrane, derived from 14 holograms, was $5 \cdot 10^{-5}$ Ns/m. In contrast, the entire membrane is more damped: the average value of R (8 holograms) was $16 \cdot 10^{-5}$, i.e. about 3 times larger than for the thin membrane. The expected values of Q for the fundamental modes of vibration of the thin and entire membrane calculated by means of Eq. (6) are about 4.7 and 2.8, respectively.

The fundamental resonance of the entire membrane is expected (and observed) at approximately 1.8 kHz (see above). The c-cells have been found to respond at about 1.5 kHz (see paper I), and the Q of their tuning curves is about 3. The vibrations recorded around 1.8 kHz by means of the capacitance electrode also have Q-values around 3 (Fig. 16). Thus, the degree of tuning of these receptor cells is approximately what one would expect, if they were responding to the fundamental mode of the entire membrane.

The fundamental resonance of the thin membrane is expected at approximately 3.4 kHz. Responses of the a- and b-cells have been found very near to this frequency (at 3.74 and 3.46 kHz, respectively; see paper I). In the a-cells the average value of Q is about 3, whereas in the b-cells it is about 6. From the amplitude of vibration a Q-value of 4.7 was expected, so it is not possible to use the value of Q to determine which of these groups are responding to the fundamental mode of the thin membrane. The Q-values are, however, certainly of the right order of magnitude.

The effect of friction can be regarded as a resisting force, opposing the movement of the membrane. In textbooks it is normally considered proportional to the velocity of the movement, and this assumption has also been made in Eq. (5). In practice, however, this is not always true. The experiments show that in the present case the assumption is not far from being correct: The values of R calculated from the holograms and from the capacitance electrode measurements are in the same order of magnitude as the Q values of the receptor cells. The former measurements were performed at sound levels about 40–60 dB higher than those needed for the determination of the tuning of the receptor cells.

It was mentioned above that the value of Q tells us how rapidly a vibration dies out, when the driving force stops. Similarly, the time required to reach steady-state after the onset of a driving force also depends on the damping of the system. If the damping is small, energy can be stored over a number of cycles, and the vibration gradually increases to steady-state magnitude. Consequently, the ear will be very sensitive to continuous sound around one of the resonance frequencies. On the other hand, the cost of this high sensitivity is a poor time resolution. The sound signals of grasshoppers consist of short, pulsed sounds with broad frequency spectra (Dumortier, 1963). Thus, the ear must be able to respond to pulses of a few cycles duration. It can be shown (see Deutsch, 1967) that—as far as physics is concerned—a Q value of 3 is the optimum for the reception of three cycle bursts of sound. In the locust ear the time resolution of one group of receptor cells (c) is much poorer than that of the other groups. The reason for this difference is probably neural, since the c-cells also respond more slowly to long pulses than do the other groups (Michelsen, 1966).

The Q values reported for the mammalian ear differ as widely as do the opinions about their importance for time resolution in the different theories of hearing (resonance-versus travelling wave-theories). Physical measurements of the vibration of the basilar membrane show Q values around 1.6–2.5 (Békésy, 1960; Johnstone and Boyle, 1967; c.f. Tonndorf and Khanna, 1968). In contrast, measurements of the system impulse response (Möller, 1970) and of tuning curves for individual units in the cochlear nerve (Evans, 1970) indicate Q-values about 10 times larger than those determined by means of physical methods.

It was mentioned in the introduction that an important argument against the occurrence of resonances in the vertebrate ear is the existence of a sharp frequency discrimination together with a fine time resolution in the same sound receptor. The weight of this argument depends upon how much of the frequency analysis is due to mechanical factors (see Möller, 1970). If a substantial part of the frequency discrimination is due to lateral inhibition, then the time resolution should not be compared with the frequency discrimination of the intact auditory system, but only with the part due to mechanical factors. It is interesting that the mechanical tuning of the basilar membrane—as measured with physical methods—is not sharper than that of the locust ear (see above). Since the locust ear is a true resonance system, it would be interesting to compare the time resolution, which has been obtained in this ear, with that of vertebrate ears. Such studies are in progress.

3. The Selectivity of the a- and b-Cells

The sensitivity of the a- and b-cells is a maximum at 3.74 and 3.46 kHz, respectively (paper I: table). In the frequency range 3–4 kHz a very dramatic change in the vibration pattern can be observed on the holograms (Fig. 13). Around 3 kHz the vibration is still governed by the entire-membrane vibration (Fig. 15), but at high intensities (Fig. 13), is can be seen to co-exist with the fundamental mode of the thin membrane (FTM). Note, that at 3 kHz the FTM is almost at the centre of the tympanum. Fig. 13 shows what happens, as the frequency is increased:

At 3.25 kHz the FTM is the dominating vibration. At 3.5 kHz almost the same picture as at 3.25 kHz is observed, but the amplitude of membrane displacement has decreased somewhat. The second mode of vibration of the entire membrane (SEM) is apparent on the thick part of the tympanum. At 3.75 kHz the FTM is more concentrated, and the amplitude of the SEM has increased (Fig. 13). At 4 kHz approximately the same picture is seen, but the amplitude of the FTM has decreased further. At higher frequencies the amplitude of vibration decreases even further, and the centre of the thin membrane vibration moves towards the centre of the membrane.

The a- and b-cells are attached to the elevated process and the styliform body, respectively. The space between these cuticular bodies is filled with a thin mass of cells, and the areas of attachment are also very close to each other. It is therefore not surprising that the best frequencies are very similar. The difference between their best frequencies was just significant ($P = 98\%$), but the difference for other response parameters was highly significant (paper I, table).

The Q-factor of the b-cells is about two times greater than that of the a-cells. Nevertheless, the a-cells are about 5 dB more sensitive than the b-cells. This is surprising, since—assuming that both responses were due to fundamental modes—a sharper tuning would mean a smaller R and thus a larger amplitude of vibration [see Eq. (10)]. Therefore, from their tuning one would expect the b-cells to be about 6 dB more sensitive than the a-cells. The experimental results, however, show that the opposite is true. Furthermore, the best frequency of the b-cells (3.46 kHz) is closer to the resonance of the FTM than is the best frequency of the a-cells (3.74 kHz). Here again, one would expect the b-cells to be more sensitive than the a-cells.

The holograms (Fig. 13) show a very concentrated vibration in the a-cell area of attachment at frequencies around 3.75 kHz. Thus, a probable explanation for the large sensitivity of the a-cells is that they are situated at the centre of this vibration. If this is true, then the attachment area of the b-cells is some hundred μm from the centre of vibration. Thus, the relative sensitivities become understandable if one assumes that the a-cells are responding to the combined effect of the FTM and the outer "bubble" of the SEM, and that the b-cells respond to the FTM only.

It should be emphasized, however, that the experimental results obtained so far do not tell us whether this explanation is correct. The spatial resolution on the holograms is not good enough to allow an accurate estimate of the position of the 3.75 kHz vibration. The results obtained with the capacitance electrode in this area vary rather much, but most amplitude curves resemble that shown on Fig. 16, *III*. Unfortunately, the tip of the capacitance electrode is too large to allow the fine details in the vibration of these adjoining attachment areas to be resolved (the more sensitive oscillator mentioned above will probably allow smaller electrodes to be used in future experiments).

4. The c-Cells: Interaction of the Fundamental Modes

The c-cells may have up to three frequencies of maximum sensitivity: 1.5 kHz, 2.5–3 kHz, and 8 kHz (see paper I, Fig. 8). The response at 8 kHz is probably due to the second mode of the thin membrane (see above). The response at 1.5 kHz is sometimes absent. All cells, however, respond maximally around 2.5–3 kHz, but this response does not correspond to any of the expected resonances. In the following, it will be shown that the response at 1.5 kHz is due to the fundamental mode of the entire membrane, and that the response at 2.5–3 kHz is caused by the interaction between the fundamental modes of the entire and thin part of the tympanum.

The vibrations of the entire tympanum seem to have their centers in the thick-membrane-end of the tympanum (see above). No receptor cells attach directly to this area, but the attachment area of the c-cells (the folded body) forms a part of the boundary between the thin and the thick parts of the tympanum (Fig. 6). Thus, it would not be surprising if the c-cells were specialized in picking up some of these vibrations. It was shown above that the degree of tuning of the c-cells to 1.5 kHz is approximately what one would expect, if they were responding to the fundamental mode of the entire tympanum.

When the capacitance electrode is placed close to the folded body, the variation in the relative amplitude of vibration in the frequency range 1 to 4 kHz may be an almost true copy of the response curve of the c-cells. Furthermore, different types of vibration have been found in different recordings; these types include an almost pure 1.8 kHz response with very little tendency towards a maximum at 2.5–3 kHz (Fig. 16, I), a 3.4 kHz response with only a small tendency towards a maximum at 1.8 kHz (Fig. 16, III), and a gradual series of intermediate types (Fig. 16, II illustrates a typical example). All these types of vibration have also been observed in the response of single c-cells (paper I). The most common type of response in the c-cells were identical to type II or intermediate between type II and type III (see paper I, Fig. 7), but cell responses like type I or III have occasionally been observed.

Thus, it is possible to record vibration patterns which correspond to the observed patterns of frequency sensitivity in the receptor cells. The nature of these vibrations can be seen from the phase relationships. In the 1.8 kHz response (Fig. 16, I) the variation of the phase lag (from driving force to membrane displacement, see above) is almost equal to

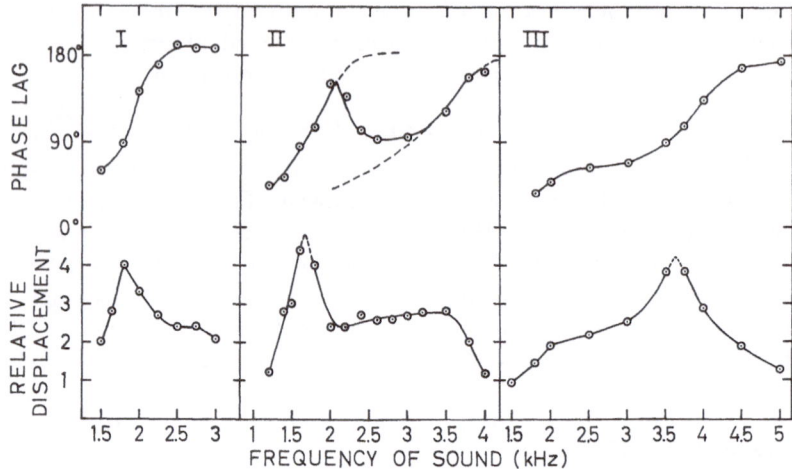

Fig. 16. Variation in amplitude and phase lag measured with the capacitance electrode placed close to the folded body (c-cells). *I* and *III* are extreme types which almost correspond to the fundamental modes of the entire and thin membrane, respectively. *II* is an intermediate type. Further explanation in the text

that expected for the resonance of a fundamental mode (see above): The phase lag is about 90° at the frequency of maximum vibration, and the curve has approximately the slope and shape expected. Also, the upper values are around 180°. Thus, the amplitude- and phase-measurements support the conclusion already reached from calculation, tuning, and holography that the 1.8 kHz response is due to the fundamental mode of the entire membrane.

From the calculations one should expect the next resonance above 1.8 kHz to be the fundamental vibration of the thin part of the tympanum at 3.4 kHz. Consequently, at this frequency the parts of the tympanum near to the centre of the thin-membrane-vibrations should have a phase lag of about 90°. In some recordings close to the folded body (Fig. 16, *III*) this is almost true. The tympanal membrane in the thick-membrane-end, on the other hand, should probably be more influenced by the entire-membrane-vibrations and thus have a phase lag of about 230° at 3.4 kHz. The membrane between the two centers of vibration can be expected to have intermediate phase lags (see above).

In principle, this is confirmed by the capacitance electrode measurements: In the recordings shown in Fig. 16, *II*, the variation in phase lag in the frequency range 1 to 4 kHz is approximately what might be expected for a point situated between the two centers of vibration. At low frequencies (around 1.8 kHz) the variation of phase and amplitude is governed by the fundamental mode of the entire tympanum (cf. Fig. 16, *I*). Around 3.4 kHz the behaviour is mainly determined by the fundamental mode of the thin membrane (cf. Fig. 16, *III*). Between these frequencies, a dramatic change of phase is observed, corresponding to a gradual shift in the relative "influence" of the two vibrations on this part of the tympanum.

During the interaction in the frequency range 2–3 kHz, the amplitudes of the two individual vibrations are added to produce a fairly constant amplitude of vibration (Fig. 16, *II*). It is interesting that the 3.4 kHz vibration does not produce any peak in the amplitude. Also, the amplitude decreases markedly around 4 kHz. At this frequency the fundamental mode of the thin membrane is interacting with the second mode of the entire tympanum (see above: a- and b-cells). This interaction seems to produce a vigorous vibration of the a-cell area of attachment (elevated process), but a decrease of the vibration amplitude at the folded body (c-cell area of attachment).

Thus, the addition of vibrations from the two sets may lead to both larger and smaller amplitudes than those expected for the individual vibrations. This is not surprising, since the amplitude of the resultant vibration depends on both the amplitudes of and the phase difference between the two components. In a second-mode-vibration the membrane on the two sides of the nodal circle is vibrating totally out of phase. The addition of a fundamental mode may therefore produce quite opposite effects with respect to resultant amplitude. Unfortunately, from the present recordings it is not clear where the nodal line of the second mode of the entire tympanum is situated on the membrane. Further studies are needed on this problem.

5. The d-Cells and the Position of the Third Mode

The amplitude- and phase-relationships illustrated in Fig. 17, left, were recorded in most preparations, when the capacitance electrode was placed close to the attachment point of the d-cells (the pyriform vesicle, see Fig. 6). The amplitudes are a maximum at 13 kHz, i.e. at the expected resonance frequency of the third mode of the thin membrane. This frequency is near the best frequency of the d-cells (12 kHz, range 10–14 kHz). The amplitude is a minimum around 8 kHz (resonance frequency of the second mode of the thin membrane). In this frequency range the d-cells are rather insensitive (see paper I: Fig. 10). The reason for this behaviour is that the area of attachment of the d-cells is close to the nodal circle of the second node at resonance. This can be seen, if the capacitance electrode is moved to other positions near the pyriform vesicle. The position of the nodal circle is also in reasonable agreement with the position expected from Eq. (6).

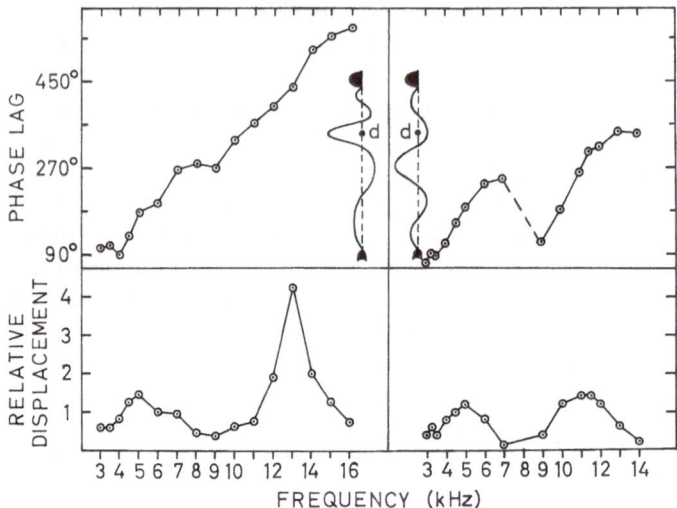

Fig. 17. Amplitude and phase measured with the capacitance electrode close to the pyriform vesicle (d-cells). *Left:* the normal case; the third mode (at 13 kHz) is so askew that its centre is at the d-cell area of attachment. *Right:* a more symmetric vibration; the pyriform vesicle is not at the centre of vibration. Note the decrease in phase at the passing of the nodal circle

It is seen from Fig. 2 that in the ideal membrane this position would correspond roughly to a maximum of vibration at the third mode, namely to the part of the membrane between the two nodal circles of the third mode. The phase-relationships, however, show that this is not true (Fig. 17, left): At 13 kHz the displacement is delayed about 90° with respect to the driving force, and not 270° as one would expect (see Fig. 2). Also, it is apparent from Fig. 17 that a 360° phase-shift is observed when the frequency is varied from about 3 kHz (approximately first mode of the thin membrane) to 13 kHz (third mode). This result can only be understood, if the d-cells respond to the movements of the centre of the third mode (and not to the "bubble" between the two nodal circles), see Fig. 17. Thus, the position of the third mode is very asymmetric. Although the technical quality of the holograms was not too good in the frequency range around 13 kHz, they do show that the maximum of displacement is centered approximately at the attachment area of the d-cells (Fig. 14).

The position of the area of attachment of the d-cells is unique, since normally the d-cells avoid responding to the second mode by being near the nodal circle at resonance, but at the same time they pick up the largest amplitude of the third mode by attaching to its centre and not to the membrane between the nodal circles.

In a few preparations, however, the third mode was probably not so askew. In these cases (Fig. 17, right) the d-cells were picking up the middle "bubble" of the third mode instead of the centre, and the amplitude of vibration was relatively smaller (assuming the 5 kHz response to be constant). Note the decrease in phase, which indicates the passing of the inner nodal circle.

Thus, in electrophysiological recordings one should expect to find a reduced sensitivity to high frequencies in some of the preparations. This was in fact observed by Popov (personal communication, 1968), who noticed that the low-frequency optimum was much more stable than the high-frequency optimum in recordings from semi-isolated ears. The difference in physical parameters, causing the different vibrations in the two groups of preparations, remains to be determined.

The fundamental mode of the thin membrane (at 3.4 kHz) is centered at the a-cell area of attachment (see above). Apparently, this vibration is so localized that it hardly affects the d-cells. On the other hand, the pyriform vesicle of the d-cells is set into vibration by the 5 kHz vibration. The nature of this vibration will be discussed below.

Discussion

1. The Mechanism of Frequency Discrimination

These results show that one should expect the thin part of the tympanum and the entire tympanum to resonate at certain frequencies (Fig. 8). Direct observations of the vibration patterns show that the actual vibrations behave almost as expected. The spatial position of the vibrations is, however, different for the two sets of vibrations. Furthermore, the positions are not constant.

In paper I it was shown that the four groups of receptor cells in the locust ear have different frequency sensitivities. The length of the sound pulses used in these experiments was 100 msec. From the damping of the tympanal membrane one may expect steady-state vibration to be reached within a few msec. Thus, it is reasonable to compare the sensitivities measured in the receptor cells with the vibration of tympanal membranes exposed to continuous sound. The use of relatively high sound intensities in laser holography and measurements with the capacitance electrode was permissible, since both the compliance and friction seem to be linear in the range of vibration amplitudes considered (see above). The selectivity of the receptor cells may be explained as follows:

The a-cells respond mainly to frequencies around 3.7 kHz. At their best frequency the basic mode of vibration of the thin membrane interacts with the second mode of the entire tympanum. As a result, the former mode concentrates around the attachment area of the a-cells

(Fig. 13). The a-cells may also respond to the second mode of the thin membrane (8 kHz).

The b-cells respond to the fundamental mode of vibration of the thin membrane (3.4 kHz). Their attachment area is some hundred μm from that of the a-cells, and they are probably not subjected to the most vigorous vibration of the first mode. They are therefore not as sensitive as the a-cells. Some b-cells also have a small second maximum around 5 kHz. The nature of the 5 kHz vibration is discussed below.

The c-cells generally respond around 2–3 kHz. Some c-cells also respond to the fundamental mode of either the entire tympanum (at 1.8 kHz) or the thin membrane (at 3.4 kHz), but others do not. The response of the c-cells seems to be determined by the interaction of the two fundamental modes of vibration (Fig. 16). Some c-cells may also respond to the second mode of the thin part of the tympanal membrane (8 kHz).

The d-cells respond predominantly to the third (13 kHz) and fourth (18 kHz) mode of vibration of the thin part of the membrane. Their attachment area on the thin membrane is near the nodal circle at resonance of the second mode, but in most preparations it is at the centre of the third mode (Fig. 17, left).

It can be concluded that the frequency discrimination in the locust ear is a purely physical phenomenon, based partly on the presence of the two sets of vibrations, and partly on the different anatomical position of the groups of receptor cells. The strongest stimulation of the receptor cells is found at the frequencies of maximum vibration of their areas of attachment on the tympanum. Normally, these frequencies are the resonance frequencies, but because of the interaction between the vibrations, this is not always true (a- and c-cells). Apparently, the selectivity of the groups of receptor cells has three different causes:

The vibrations of the entire tympanum have their centers in the thick-membrane-end of the tympanum. Consequently, only the group of cells which are attached to this area (the c-cells) respond directly to some of these vibrations. Although most of these resonances do not stimulate the receptor cells directly, they may still play an important role as "modulators" for the "dominating" thin-membrane vibrations.

A probable result of this interaction is the change of spatial position observed both in the entire-membrane vibrations and in the vibrations of the thin membrane system. The most surprising feature here is that the vibrations are extremely localized. For example, the spatially concentrated first mode of the thin membrane, which causes a vigorous movement of the a-cell attachment area (Fig. 13), hardly affects the attachment areas of the three other groups.

Finally, the groups of receptor cells may have their attachment area near one or more nodal circles (at the corresponding resonances). This principle is well illustrated in the d-cells (see above). The interaction between two vibrations may also create an "artificial nodal line", if the two vibrations oppose each other. This may be the reason for the decrease in the response of the c-cells around 4 kHz (see above).

Thus, according to this view the frequency sensitivity of a receptor cell is determined by the vibration of its area of attachment on the membrane, and not by any frequency preference of the receptor cell itself. The dynamic properties of the receptor cells, on the other hand, are probably of neural origin (cf. Nakajima and Onodera, 1969). In the locust ear two different types of adaptation have been found (Michelsen, 1966); the c-cells respond more slowly than the other cells. It is remarkable that, although the attachment of the c-cells to a more damped membrane should, in theory, allow them to have a better time resolution than the other cells, the opposite is in fact true.

During the recordings from single receptor cells an abnormal unit was found in the anatomical region of the c-cells. The frequency sensitivity of this cell was similar to that of a typical c-cell (paper I, Fig. 7, left), but it had an a-cell type of adaptation. This observation can be explained by assuming that, during the ontogeny, an a-cell had attached itself to the c-cell part of the tympanum. Thus, the frequency sensitivity would be determined by the area of attachment, whereas the adaptation would remain identical to the a-cell type.

The frequency discrimination in the locust ear is not very sharp compared to that of vertebrate ears. It should be noted, however, that the vibration of the tympanum is so complex above 3 kHz that it is possible—for a given sound frequency—to find a point on the membrane which has a maximum of vibration at that frequency. Thus, a much better frequency discrimination might have been obtained, if more groups of receptor cells had attached to the tympanum.

2. Some Unsolved Problems

The present study can only be regarded as the first step towards an understanding of the mechanism of frequency discrimination in the locust ear. A number of uncertain points and unsolved problems have been neglected in the main part of the paper in order not to make the description more complicated than necessary. Two problems will be discussed here: Are there other vibrations than those expected for the ideal membrane? And, to what kind of vibration are the receptor cells subjected?

Irregular Vibrations. In general, the vibrations of the tympanum can be regarded as composed of two sets of circularly symmetric modes. It was mentioned in the introduction that, in an ideal membrane acted

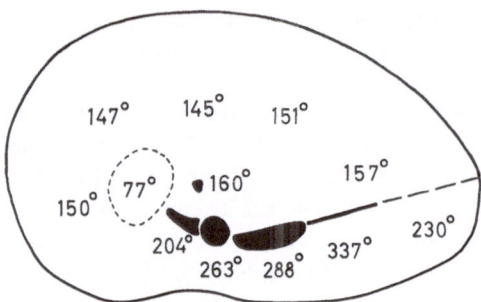

Fig. 18. The variation in phase lag over the membrane at 5 kHz. The values are the average of about 150 measurements from 5 different preparations. Further explanation in the text

upon by a uniform force (sound wave), only the circularly symmetric vibration patterns can be expected. The tympanal membrane is, however, not homogeneous or circular, so it is possible that other types of vibration occur. In the present study at least one vibration was observed, which does not fit into the calculated series of vibrations.

At 9.5 kHz an unexpected vibration has been observed on the holograms in the area between the centre of the membrane, the folded body (c-cells), and the pyriform vesicle (d-cells), see Fig. 15. It can also be recorded by means of the capacitance electrode. The phase lag at the 9.5 kHz peak was about 270°. Apparently, this vibration is restricted to a very small area of the membrane, and it does not seem to affect the receptor cells. At the frequencies around 9.5 kHz several irregular vibrations can be expected (see Morse, 1948). The nature of this vibration remains unknown.

In most recordings with the capacitance electrode from the thin membrane one of the maxima in the amplitude of vibration was found around 5 kHz (Fig. 17 shows typical examples). Also, some b-cells have a small second sensitivity maximum around 5 kHz (paper I, Fig. 6). At this frequency the thin-membrane system should be changing from the first (3.4 kHz) to the second (8 kHz) mode, and the entire-membrane system should be between the second (4.2 kHz) and the third (6.6 kHz) mode. Thus, the amplitude of vibration should be relatively small in both systems. Nevertheless, an amplitude maximum is often observed.

The holographic pictures made at 5 kHz differ very much. One of them was almost identical to the 4 kHz vibration shown in Fig. 13; i.e. the centers of vibration were at the a-cell area and at the thick membrane, respectively. Another picture is shown in Fig. 15. In this case the centers are nearer to the geometrical center of the tympanum. The amplitude of the 5 kHz vibration on Fig. 15 is approximately what one would expect from the computed values (Fig. 9).

The phase lags observed at 5 kHz differ rather much over the membrane (Fig. 18). At the thin membrane values around 150° are observed, whereas the thick membrane has phase lags up to about 340°. These values are the average of about 150 measurements on 5 preparations, and they are near to the expected phase lags (170 and 340°, respectively). In two preparations an area with phase lags around 80° was observed on the thin membrane near to the b-cells.

Thus, both the amplitude and phase lag are near to the expected values for the circularly symmetrical vibrations. The only unexpected finding is the area of 80° phase lag. The most probable explanation for the observed amplitude maximum and the response of the b-cells is that the centre of the thin-membrane vibration (which at 3–4 kHz is at the a-cell area) is moving towards the centre of the tympanum at 5 kHz. Therefore, the capacitance electrode (and the receptor cells attaching to the thin membrane) will record a maximum of amplitude, although the thin-membrane vibration should be a minimum at 5 kHz!

The only resonance expected for the thin-membrane system around 5 kHz is that of the simplest irregular vibration (d on Fig. 1) at 5.4 kHz (see Morse, 1948). In this vibration there should be a nodal line across the membrane, and the membrane on each side of this nodal line should vibrate out of phase. The addition of a weak "d-vibration" and the circularly symmetric modes might give a vibration pattern as that seen on Fig. 15. This "explanation" is, however, only a guess. Further observations are needed to solve the problem.

Excitation of the Receptor Cells. The technique used here was sufficient to demonstrate the presence of the vibration patterns of the membrane, but it did not allow a detailed study of the vibration of the attachment areas. It may, however, be useful to speculate about some of the problems which remain to be solved in future experiments:

In the above description the excitation of the receptor cells has been said to be determined by the vibration of their attachment areas on the membrane. In general, this may be true, but it should be borne in mind that the cell bodies are situated in a mass of cells, Müller's organ (Fig. 6). The anatomy of the receptor cells and their (indirect) attachment to the cuticular bodies are extremely complex (see Gray, 1960). The excitation of the receptor cells is likely to be determined by the displacement of the dendrites *relative* to the rest of the cells. Therefore, it is important to know whether the mass of cells in Müller's organ can be set into vibration by the movements of the tympanum.

The attachment parts of the tympanum have been referred to as "areas", but (except for the pyriform vesicle) they are certainly not simple thickenings of the cuticle (see Schwabe, 1906). The styliform body has the shape of an hour-glass, which is fastened to a plate at the membrane-end. The shape of the hollow elevated process is often described as that of a cup, but in fact it bends at some distance from the tympanal membrane. Finally, the folded body is a surprisingly large and complicated, folded thickening of the membrane. It would not be surprising if these complicated structures have a function in the transmission of the vibrations to the receptor cells.

I should like to express my most sincere thanks to O. Juhl Pedersen M. Sc. and Knud Rasmussen M. Sc. (The acoustics Laboratory, The technical University of Denmark) for their invaluable help on the physical problems. Without their continuous advice during the years this investigation could not have been carried out.

I am most grateful to Professors Bent Christensen, Franz Huber, and Ulrik Lassen for discussions and excellent working conditions; to Hans Andersson, Hans Bjelkhagen, and Docent Nils Abramson for carrying out the laser experiment; to Professor E. B. Hansen and Mr. Kent Hansen for help with the handling of Eqs. (5) and (6); and to McGraw-Hill Book Company for permission to reproduce Fig. 1.

It is a pleasure for me to acknowledge the generous support from the Deutsche Forschungsgemeinschaft, the Carlsberg Foundation, and Statens naturvidenskabelige Forskningsraad.

Appendix

The Radiation Impedance

The vibration of the tympanum, caused by a given sound wave, is determined mainly by the properties of the membrane. However, the air will react against the movement. The contribution of this interaction can be measured as a radiation impedance. At "low" frequencies the radiation impedance may be represented by a combination of a mass and a frequency-dependent resistance. Therefore, interactions between the membrane and the air will affect both the resonance frequencies and amplitude of vibration. An exact calculation of the radiation impedance is difficult. For the circular plate surrounded by an infinite baffle the radiation impedance for different modes of vibration have been calculated by Lax (1944). A similar approach can be made for the membrane, but in the present case the membrane is unbaffled, and no exact formula exist for this case. However, the order of magnitude of the radiation impedance can be estimated by means of some simple models.

The radiation impedance (Z_r) of the unbaffled membrane may be estimated by computing that of a spherical dipole source with the same radius as the membrane. It can be shown (see Morse, 1948) that

$$Z_r = R_r + j X_r \simeq \frac{16 \cdot \pi^5 \cdot \varrho \cdot f^4 \cdot a^6}{3 c^3} + j \cdot \omega \cdot \frac{2\pi \cdot \varrho \cdot a^3}{3} \tag{11}$$

where R_r = radiation resistance (Ns/m),
X_r = reactance component of Z_r (Ns/m) = $\omega\, m_r$, where
m_r = radiation mass (kg),
$j = \sqrt{-1}$,
$\pi = 3.14159\ldots$,
f = frequency of sound (s^{-1}),
a = radius of membrane (m),
c = velocity of sound (approximately 344 m/s),
ϱ = density of air (approximately 1.2 kg/m^3),
ω = angular frequency ($= 2\pi f$).

It is apparent from Eq. (11) that the reactive component of the radiation impedance is given by half the mass of a volume of air equal to that of the equivalent dipole source. In the present case one finds $m_r = 2.5\ \mu$g for $a = 1$ mm.

The value calculated by means of the spherical dipole-model may be compared with the value estimated for a plane, unbaffled, circular disk moving as a rigid piston. It can be shown (Wiener, 1951; Beranek, 1954) that in the size and frequency range considered here,

$$m_r = 0.85\pi \varrho a^3. \tag{12}$$

From this equation one finds $m_r = 3.2\ \mu g$ for $a = 1$ mm. This result is fairly close to that found above (2.5 μg). When compared with the weight of the thin part of the tympanum (about 9.2 μg), the radiation mass is far from negligible.

The radiation mass calculated by means of these models may be used for the first mode, where all parts of the membrane vibrate in phase. At present no formula exist for computing the exact radiation mass at higher modes of vibration of the unbaffled membrane, but it seems safe to assume that it will be considerably smaller than that for the first mode. If one considers a spherical sound source of n-th order at "low" frequencies ($\omega\, a/c < 1$, i.e. $f < 50$ kHz for $a = 1$ mm), the radiation mass is approximately given by (Morse, 1948)

$$m_r = \frac{3\,M}{(n+1)\,(n+2)} \tag{13}$$

where $M = \varrho\, 4\pi a^3/3 = $ the mass of the equivalent volume of air. Using this model and $a = 1$ mm, the radiation mass becomes 2.5, 1.0, 0.5, and 0.3 μg for the first four modes, respectively [for $n = 1$ the expression is identical to Eq. (11)]. Approximately the same reduction was found by Lax (1944) for plates surrounded by infinite baffles. Although these models are not valid for the present case, the same degree of reduction has been used in the calculations above. The error introduced is minimal.

The mass per unit area (σ) is estimated from the mass of the membrane plus the radiation mass. Since the latter depends upon the order of the vibrational mode, the total effective mass must be calculated for each mode. It was mentioned above that the membrane behaves as a simple, driven oscillator at frequencies up to the first resonance frequency, but that certain corrections are necessary in order to fit the measured membrane parameters to the equations for simple oscillators. If one calculates the equivalent membrane impedance for "low" frequencies (see Morse, 1948), one finds an effective mass of 1.38 times the total mass of the membrane [and an effective stiffness of 8π times the membrane tension, cf. Eq. (4)]. Therefore, when the real mass and the radiation mass were added, the latter was reduced by 1/1.38. It is not quite clear whether this reduction is permissible also for higher modes, but here again the error introduced is minimal.

From formula (11) it is seen that whereas the resistive component (R) of the membrane impedance (Z) is normally considered to be independent of frequency, the resistive component (R_r) of the radiation impedance (Z_r) increases with the fourth power of frequency. Using $a = 1$ mm and $f = 10$ kHz, one finds $R_r = 5 \cdot 10^{-7}$ mks mechanical ohms (Ns/m). At 20 kHz R_r becomes $8 \cdot 10^{-6}$ Ns/m.

The magnitude of R_r derived for a plane, unbaffled, circular disk moving as a rigid piston is given (Beranek, 1954) by

$$R_r = 0.19\, a^6\, \varrho\, \omega^4/c^3. \tag{14}$$

The value obtained by means of this equation is about 5 times smaller than that calculated from Eq. (11).

It has been shown above that the resistive R of the membrane is about $5 \cdot 10^{-5}$ Ns/m for the thin membrane (where $a = 0.9$ mm) and $16 \cdot 10^{-5}$ Ns/m for the entire tympanum (where $a = 1$ mm). Thus, even at 20 kHz (where the thin membrane is vibrating in its fourth mode, and the value of R_r therefore probably should be reduced, see Lax, 1944) the magnitude of R_r is at least 10 times smaller than R. For the present purpose it therefore seems safe to neglect the radiation resistance.

References

Abramson, N., Andersson, H., Bjelkhagen, H.: Hologram avslöjar farlige flygplans-
fel. Ny Teknik (Stockh.) 1970, 11, 3–5.

Autrum, H.: Anatomy and physiology of sound receptors in invertebrates. In:
Acoustic behaviour of animals (R.-G. Busnel, ed.), p. 412–433. Amsterdam:
Elsevier 1963.

Békésy, G. von: Experiments in hearing. New York: McGraw-Hill 1960.

— Resonances in the cochlea? Sound 3 (4), 86–91 (1969).

— Travelling waves as frequency analysers in the cochlea. Nature (Lond.) 225,
1207–1209 (1970).

Beranek, Leo L.: Acoustics. New York: McGraw-Hill 1954.

Crandall, Irving B.: Theory of vibrating systems and sound. London: MacMillan
1927.

Deutsch, S.: Models of the nervous system. New York: Wiley 1967.

Dumortier, B.: The physical characteristics of sound emission in Arthropoda. In:
Acoustic behaviour of animals (R.-G. Busnel, ed.), p. 346–373. Amsterdam:
Elsevier 1963.

Evans, E. F.: Narrow "tuning" of cochlear nerve fibre responses in the guinea-pig.
J. Physiol. (Lond.) 206, 14–15 P (1970).

Gabor, D., Stroke, G. W.: Holography and its applications. Endeavour 28, 40–47
(1969).

Gray, E. G.: The fine structure of the insect ear. Phil. Trans. B 243, 75–94 (1960).

Helmholtz, H. von: Die Lehre von den Tonempfindungen als physiologische Grund-
lage für die Theorie der Musik. Braunschweig: Vieweg 1862.

Herzog, R. O.: Fortschritte in der Erkenntnis der Faserstoffe. Z. angew. Chem. 39,
297–302 (1926).

Huxley, A. F.: Is resonance possible in the cochlea after all? Nature (Lond.) 221,
935–940 (1969).

Jensen, M., Weis-Fogh, T.: Biology and physics of locust flight. V. Strength and
elasticity of locust cuticle. Phil. Trans. B 245, 137–169 (1962).

Johnstone, B. M., Boyle, A. J. F.: Basilar membrane vibration examined with the
Mössbauer technique. Science 158, 389–390 (1967).

Lax, Melvin: The effect of radiation on the vibrations of a circular diaphragm.
J. acoust. Soc. Amer. 16, 5–13 (1944).

Michelsen, Axel: Pitch discrimination in the locust ear: observations on single
sense cells. J. Insect Physiol. 12, 1119–1131 (1966).

— The physiology of the locust ear. I. Frequency sensitivity of single cells in the
isolated ear. Z. vergl. Physiol. 71, 49–62 (1971).

Möller, Aage R.: Studies of the damped oscillatory response of the auditory
frequency analyzer. Acta physiol. scand. 78, 299–314 (1970).

Morse, Philip M.: Vibration and sound, 2. ed. New York: McGraw-Hill 1948.

Nakajima, S., Onodera, K.: Membrane properties of the stretch receptor neurones
of crayfish with particular reference to mechanisms of sensory adaptation.
J. Physiol. (Lond.) 200, 161–185 (1969).

— — Adaptation of the generator potential in the crayfish stretch receptors under
constant length and constant tension. J. Physiol. (Lond.) 200, 187–204 (1969).

Pennington, K. S.: Advances in holography. Scient. Amer. 218, 40–48 (Febr. 1968).

Powell, R. L., Stetson, K. A.: Interferometric vibration analysis by wavefront
reconstruction. J. opt. Soc. Amer. 55, 1593–1598 (1965).

Pumphrey, R. J.: Hearing in insects. Biol. Rev. 15, 107–132 (1940).

Rayleigh, Lord: Theory of sound. London: MacMillan 1926.

Schwabe, J.: Beiträge zur Morphologie und Histologie der tympanalen Sinnes-apparate der Orthopteren. Zoologica **20**, 1–154 (1906).

Skudrzyk, E.: Die Grundlagen der Akustik. Wien: Springer 1954.

Stetson, K. A.: Vibration measurement by holography. In: Symposion on the engineering uses of holography (E. R. Robertson, ed.), p. 307–331. Cambridge: University Press 1970.

Tonndorf, J., Khanna, S. M.: Displacement pattern of the basilar membrane: a comparison of experimental data. Science **160**, 1139–1140 (1968).

Whitfield, I. C.: Coding in the auditory nervous system. Nature (Lond.) **213**, 756–760 (1967).

Wiener, F. M.: On the relation between the sound fields radiated and diffracted by plane obstacles. J. acoust. Soc. Amer. **23**, 697–700 (1951).

Axel Michelsen
Zoological Laboratory
Universitetsparken 15
DK-2100 Copenhagen O, Denmark

Z. vergl. Physiologie 71, 102–128 (1971)
© by Springer-Verlag 1971

The Physiology of the Locust Ear

III. Acoustical Properties of the Intact Ear

Axel Michelsen

Laboratories of Zoology and Zoophysiology B, University of Copenhagen, and
Zoologisches Institut, Lehrstuhl für Tierphysiologie, Universität zu Köln

Received October 1, 1970

Summary. The sensitivity of three different preparations of the tympanal organ ("isolated", "operated", and "intact", see Fig. 7a–c) has been measured over a wide range of frequencies (Figs. 3 and 6). The sensitivity of the intact ear to low frequency sound depends on the fat content of the animal (Figs. 4 and 5). The effect of diffraction (Fig. 8), the sound absorption in internal tissues (Fig. 10), and the sound transmission through the animal (Fig. 9) have been measured in order to explain the observed sensitivities in the three preparations. The internal tissues seem to act as an acoustic low-pass filter. Therefore, at high frequencies the intact and operated ears are acting almost as pressure receivers (Fig. 2a). The isolated ear is acting as an unbaffled pressure gradient receiver (Fig. 2b) with an "effective distance" of 0.8 mm. A mathematical model for asymmetric sound receivers is presented and used to calculate the force acting to move the tympanum in the operated ear at low frequencies. The driving forces in intact and operated ears are of the same order of magnitude as in a similar pressure receiver (table). The membrane vibrations at high frequencies are heavily damped both by the radiation resistance and by friction in the internal tissues behind the ear. The implications of these results for the understanding of directivity are discussed. Some common methods for determination of threshold are compared.

Introduction

For 30 years insect ears (tympanal organs) have been considered almost pure pressure gradient receivers (Pumphrey, 1940; Autrum, 1941). The pressure gradient is a vectorial component of sound, and therefore each tympanal organ should be a directional instrument, capable of detecting the direction of sound waves. One aim of the present study is to investigate to what extent these ideas are valid for the intact ear of the locust. An attempt has also been made to provide a theoretical basis for studies on directivity of insect ears.

In part I of this series of papers it was shown that the four anatomical groups of receptor cells in the *isolated* locust ear have different frequency sensitivities. In part II the physical basis for the frequency discrimination was described: it was shown that two sets of selective resonances exist in the tympanal membrane; the selectivity of the receptor cells is due to their different spatial position, which allows them to pick up different resonances.

The threshold curves of *intact* locust ears differ considerably from those of isolated ears. There is also a large difference between the threshold curves of different intact ears. In this paper an attempt is made to explain these findings by considering the differences in the acoustical conditions of isolated and intact ears, and by comparing intact ears with operated ears. The physical properties of intact ears are so complicated, however, that only a tentative explanation can be given in this paper; some possible lines for further studies are nevertheless outlined, to stimulate research on the physical basis of insect hearing.

Terminology. The use of physical terms to describe the transducing mechanism of insect ears has been the subject of much confusion. The physical parameters of sound acting to move the tympanal membrane may be the pressure, the pressure gradient, or a combination of both. Terms such as "displacement-", "velocity-" and "acceleration-receptor", on the other hand, refer to the particular physical parameter of the movement which is capable of exciting the receptor cells. Obviously, these two sets of terms refer to two different problems, and the common usage of for example "displacement receptor" as synonymous with "pressure gradient receptor" should be avoided.

Methods

Adult desert locusts (*Schistocerca gregaria* Forskål, phase gregaria) were not used until at least three weeks after their final moult. Some of the animals were starved for various periods before the experiments, in order to reduce the amount of fat in the abdomen. Various operations were performed in some of the experiments. These operations will be described below.

The experimental set-up and the acoustic conditions were similar to those previously described (paper I).

The head, wings and legs were removed. The animal was pinned down on a small platform of wax with its ventral side upwards and its longitudinal axis perpendicular to the direction of sound. A part of the ventral cuticle covering the third thoracic ganglion was removed. The sensory response of the entire tympanal nerve was recorded by means of a hook electrode (50 μm platinum).

Various methods for estimating the threshold were tested. It soon became apparent that only one of the methods was accurate enough for the present purpose. Since the threshold values reported in the literature differ enormously (Fig. 13), a comparison of some of the commonly used methods is given in the appendix. The method used here is a semistatistical one:

In the absence of sound a spontaneous activity of about 380 spikes per second is observed in the entire nerve-recording (average of 22 preparations). There are about 60–70 receptor cells in the ear, and the spontaneous activity of the individual cells is about 5 spikes per second (average of 35 preparations). Thus, it is reasonable to assume that almost all the spontaneous action potentials travelling in the tympanal nerve are observed. The individual spikes last about one msec, so with this rate of firing there is not much summation (Fig. 1).

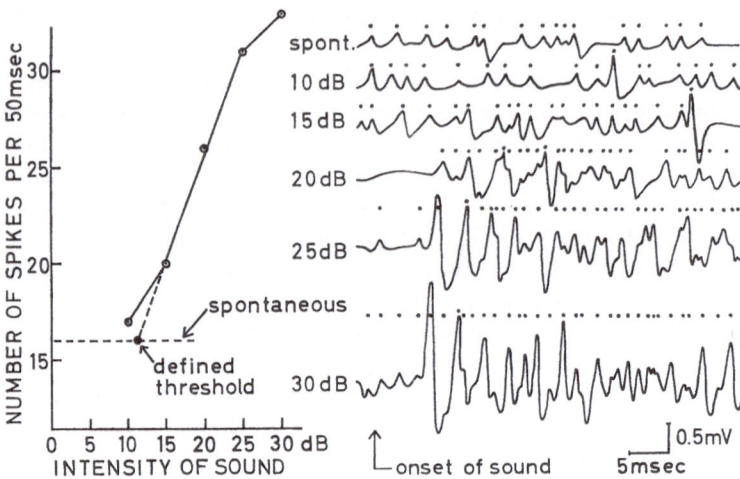

Fig. 1. The method used to determine the threshold in recordings from the entire nerve. The dots indicate the spikes counted (cf. the text)

When the intensity of sound is increased, the number of spikes increases (Fig. 1). Several of the spikes will now summate; the form of the summated spikes depends on the (random) synchronization of their components. If the total number of visually discernable spikes for a given period of time is plotted against intensity, a linear relationship over a certain intensity range is obtained in most cases. (This result is not directly obvious, because the firing rate of the individual receptor cell increases in an exponential manner). A threshold value can now be defined by extrapolation of the straight line to the level of spontaneous activity (Fig. 1).

The threshold thus defined is, of course, quite arbitrary, but it can be reproduced in successive recordings with an accuracy of ± 1–2 dB. In contrast, the accuracy of most other methods is about ± 5 dB (see the appendix). Further, this method is not biased in favour of the large groups of receptor cells as are the methods in which a certain level of summation is measured. This is important if the sensitivities to low and high frequencies are to be compared, since the number of receptor cells responding at high frequencies (the d-cells) is less than one fifth of the total number.

In practice, the response to 5 sound pulses of 45 msec duration was recorded at each intensity. The two highest values were discarded, and an average was calculated from the remaining three responses. This procedure was necessary, because periodic bursts of spikes occurred in connection with the respiratory movements. In a few preparations the respiration was so vigorous that it hindered the use of the method.

This method is simple but laborious (about 20000 spikes had to be counted for each threshold curve). An attempt was made to construct an electronic device to do the counting, but it appeared to be a difficult problem. However, it should not be impossible to develop this method to a practical but nevertheless accurate way of determining the threshold in whole nerve recordings.

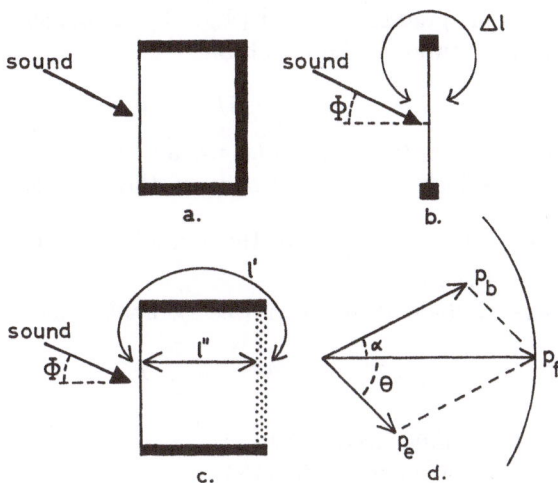

Fig. 2a–d. A pressure receiver (a), a symmetrical pressure gradient receiver (b), and an unsymmetrical sound receiver (c). In (d) the sound pressure on the front (p_f) and back (p_b) of the membrane are drawn as vectors with the phase difference α. p_e and Θ indicate the resultant equivalent pressure and phase angle. Further explanation in the text

The Force Acting on Tympanal Membranes

1. Pressure and Pressure Gradient Receivers

In many microphones the movements of a membrane are recorded by means of a suitable transducer system. Such microphones can be divided into two broad classes: pressure and pressure gradient microphones. In a pressure microphone the membrane is backed by a closed chamber, and the sound waves reach only the front of the membrane (Fig. 2a). In this case the force (F) acting to move the membrane is

$$F = A \cdot p \tag{1}$$

where $A =$ membrane area (m²),
$p =$ sound pressure (N/m²).

Since pressure is a scalar quantity, the force will be independent of the direction of the sound wave.

In a pressure gradient microphone (for example a ribbon microphone) both sides of the membrane are exposed to the sound wave (Fig. 2b). Such a device will respond to the difference in sound pressure on the front and back of the membrane. If the microphone is small enough, this difference may be represented by the pressure gradient, hence the name "pressure gradient microphone". The force acting on a small, symmetrical pressure gradient receiver (Fig. 2 b) will depend on the angle of incidence (Φ) of sound on the membrane:

$$F = - A \cdot \frac{dp}{dx} \cdot \Delta l \cdot \cos \Phi \qquad (2)$$

where $x =$ direction of travel of sound wave,

$\Delta l =$ the "effective distance" between the two sides (m).

Here, $\frac{dp}{dx} \cos \Phi$ is the component of the x gradient of pressure acting across the faces of the membrane.

The pressure gradient is given by (see Beranek, 1954)

$$\frac{dp}{dx} = -j \frac{\omega}{c} p = -j \frac{2\pi}{\lambda} p \qquad (3)$$

where $j = \sqrt{-1}$,

$\omega = 2 \pi f =$ angular frequency (s^{-1}),

$c =$ velocity of sound in air (m/s),

$\lambda = c/f =$ wavelength of sound (m),

$f =$ frequency of sound (s^{-1}).

Thus, Eq. (2) becomes

$$F = j A \frac{\omega}{c} p \Delta l \cos \Phi = j A \cdot p \cdot \frac{2 \pi \Delta l}{\lambda} \cos \Phi. \qquad (4)$$

When Eqs. (1) and (4) are compared, it is seen that in both cases the force is determined by $A \cdot p$. In the pressure receiver the force is in phase with the sound pressure. In the pressure gradient receiver, however, the force is 90° out of phase with the sound pressure (this is indicated by j); also, the force will be proportional to the phase difference between the sound acting on the front and back side of the membrane, respectively [this is indicated by the last part of Eq. (4)].

It should be noted that these equations are valid only for plane sound waves, i.e. at a relatively long distance from the sound source. Near a loudspeaker the sound waves will be more or less spherical; and in a spherical sound field the magnitude of the pressure gradient may theoretically be infinite. In practice, it is often several dB greater than in a plane sound field of identical sound pressure.

For several years it has been recognized that insect ears have many features in common with pressure gradient microphones (Autrum, 1941; Pumphrey, 1940). Thus, the directional sensitivity of insect ears has been ascribed to the variation in the angle of incidence of the sound upon the tympanal membrane [Φ of Eqs. (2) and (4)].

2. The Asymmetric Sound Receiver

The individual ear in an intact locust is not a symmetrical sound receiver comparable to that illustrated in Fig. 2 b. Therefore, in order to evaluate the function of the intact locust ear, we will have to consider the characteristics of a non-symmetrical sound receiver (Fig. 2c).

Let p_f and p_b be the sound pressures acting on the front and back sides of the membrane, respectively. The phase angle between p_f and p_b due to the "effective distance" will be given by $2\pi \Delta l/\lambda$ [cf. Eq. (4)]. If the sound passes a sound absorbing material on its way to the back of the membrane an additional phase angle (φ) may be introduced. The total phase angle (α) between p_f and p_b will now be

$$\alpha = \frac{2\pi \Delta l}{\lambda} + \varphi. \tag{5}$$

In order to calculate the magnitude and phase of the resultant force, p_f and p_b are drawn as vectors (Fig. 2d). Their difference gives the pressure p_e which in a pressure receiver would be equivalent to the force divided by the membrane area [see Eq. (1)]. Using simple trigonometrics one finds

$$p_e = \sqrt{p_f^2 + p_b^2 - 2p_f p_b \cos \alpha}. \tag{6}$$

The phase angle (Θ) between p_f and p_e is found from (Fig. 2d):

$$\cot \Theta = \frac{p_b \cos \alpha - p_f}{p_b \sin \alpha}. \tag{7}$$

If $p_f = p_b =$ the sound pressure (p) in the surrounding medium, these equations reduce to

$$p_e = p \cdot \sqrt{2} \cdot \sqrt{1 - \cos \alpha}, \tag{8}$$

$$\cot \Theta = \frac{\cos \alpha - 1}{\sin \alpha} = \cot \alpha - \csc \alpha. \tag{9}$$

In the symmetrical case (Fig. 2b) the phase difference corresponding to α was determined by the "effective distance" (Δl) together with the angle of incidence (Φ). In the asymmetric receiver (Fig. 2c) this is obviously not the case. Here, Δl may be divided into l' (the effective distance on the outside of the "box") and l'' (the effective distance inside the "box"), see Fig. 2c. Considering the contribution to α, the angle of incidence (Φ) will only affect l'. Thus, the Δl used in Eq. (5) will be given by

$$\Delta l = l' \cos \Phi + l''. \tag{10}$$

Sensitivity of Intact and Operated Ears

The threshold curves of locusts taken from a normal laboratory culture are fairly similar in shape, but the absolute sensitivity to low frequencies varies considerably (Fig. 3 shows the extreme threshold curves). At 3.5–4 kHz the variation may be about 30 dB (range 7 to

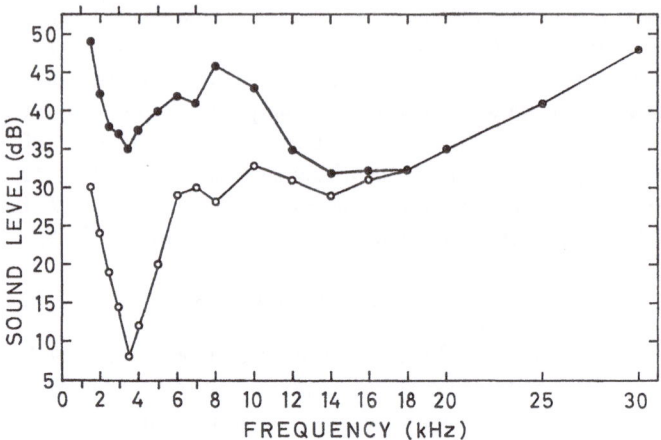

Fig. 3. The threshold (in dB re. $2 \cdot 10^{-5}$ N/m²) of an extremely meagre locust (○)
and an extremely fat locust (●)

35 dB re. $2 \cdot 10^{-5}$ N/m²). The inaccuracy in the experimental determination of the threshold was only a few dB.

In the late summer of 1968 the animals were very sensitive (threshold at 3.5–4 kHz about 7–10 dB). Dissections demonstrated a massive infection with nematodes (*Mermis sp.*) and a large reduction in the size of the fat body and ovaries. A systematic study of the relationship between the fat content and auditory sensitivity was then undertaken. The influence of the fat body on hearing has been suggested earlier by Autrum *et al.* (1960).

Influence of Fat. Fig. 4 shows the relations between the total amount of fat in the abdomen (i.e. the amount of fat which could be removed in a quick dissection) and the threshold at 3.5–4 kHz. The sensitivity is high when the fat content is small, and vice versa. There is, however, a considerable variation of the thresholds at intermediate amounts of fat. Dissections showed that this variation is correlated with differences in the distribution of fat in the body: In some cases the fat body was in the abdomen, and very little fat was found between the ears. Such preparations were fairly sensitive. In other preparations lumps of fat were attached to the tracheal air sacs between the ears. These animals were less sensitive to sound.

In the dissections it is difficult to remove only the tracheal walls of the air sacs of relevance to the hearing organs. Therefore, the sensitivity to other frequencies was plotted against the threshold at 3.5–4 kHz. Fig. 5 shows some examples; a fairly large variation is observed at most

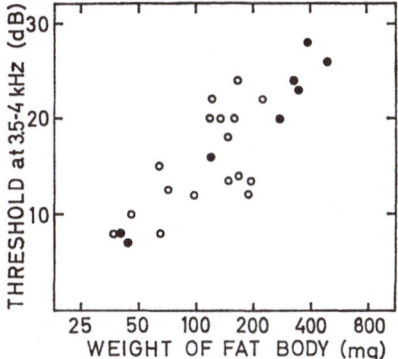

Fig. 4. The relationship between the total amount of fat body and the threshold
at 3.5–4 kHz (○ = ♂, ● = ♀)

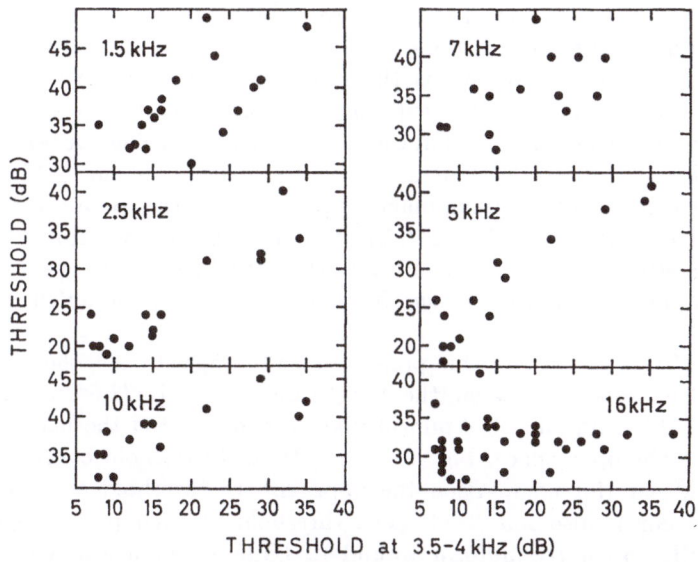

Fig. 5. The threshold at various frequencies plotted against the threshold at
3.5–4 kHz

frequencies, but the sensitivity to high frequencies (12–16 kHz) remains
almost constant.

Ovaries and Gut. The variation in sensitivity is normally correlated to
the amount of fat between the ears. In some insensitive females, however,
only little fat was present; in these cases the anterior part of the ovaries

A. Michelsen:

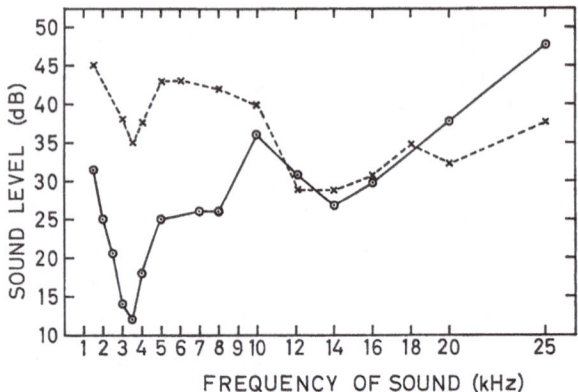

Fig. 6. The threshold (in dB re. $2 \cdot 10^{-5}$ N/m^2) for the operated ($\circ$) and the isolated ($\times$) ear preparations

was found in the region of the ears. Normally, the ovaries are situated in the abdomen rostral to the ears.

In insensitive preparations the removal of the gut was found to increase the sensitivity to low-frequency sound. By this operation an artificial free space is created in the region of the ears. The gut is situated ventrally relative to the ears, and probably it does not normally affect the hearing ability: Preliminary experiments were performed with starved animals which had a fairly uniform sensitivity. Half of these animals were given a solid meal before the experiment; the sensitivity of these animals did not differ significantly from that of the unfed control animals.

The Operated Ear. In this preparation the opposite ear, the air sacs, and other tissues between the ears were removed. Thus, a tubular channel (5–6 mm long, 3–4 mm in diameter) connected the back of the ear with the open space. Fig. 6 shows a typical threshold curve for the operated ear (for comparison the approximate threshold curve of the isolated ear is also indicated). The threshold of such preparations at 3–3.5 kHz was fairly uniform around 12 dB (average of 9 preparations, range 9–15 dB). By comparison, the average threshold at 3.5–4 kHz in the most sensitive, intact ears was 8 dB (10 preparations, range 7–10 dB). The sensitivity of the operated ear was independent of the previous sensitivity. Thus, fat animals became more sensitive when operated on, while the ears of meagre animals became about 4 dB less sensitive.

Two out of 9 operated ears were more sensitive at 3 kHz than at 3.5 kHz. In contrast, four out of 13 intact ears were more sensitive at 4 kHz than at 3.5 kHz. Thus, the operation may cause a small change of best frequency, but it is not likely to exceed 500 Hz.

The Physics of Intact and Operated Ears

1. The Effect of Diffraction

Eqs. (2) and (4) were derived on the assumption that the membrane and its surrounding structures are so small relative to the wavelength of sound that they do not disturb the sound field. In practice, this means that the diameter of the sound receiver must be less than about one tenth of the wavelength. The frequency range of the locust ear extends from 1 to 40 kHz, i.e. from a wavelength of 34 cm to one of 8 mm. The shape of a locust is approximately that of a cylinder, 5 cm long and about 0.6–0.8 cm in diameter. Thus, the locust is likely to disturb the sound field, except in the very lowest end of the frequency range of the ear.

At present, it is only possible to calculate the effect of diffraction for obstacles of very simple shape, e.g. a sphere or a cylinder of infinite length. Furthermore, the scattering of sound depends on the acoustic properties of the body. No experimental data are available for the sound-absorbing properties of total insect bodies.

The best way to measure the effect of diffraction on the sound pressure at the position of the ear is to replace the ear with a very small microphone. Ideally, the microphone should have the same acoustic impedance as the ear, but this is not possible to realize in practice. The smallest microphone at my disposal was a $1/_4''$ condenser microphone (Brüel & Kjär, 4135) mounted on an adaptor (UA 0035) connected to a cathode follower (2615). The diameter without the protecting grid was about 6 mm. This microphone (without grid) was placed in a locust with its membrane in the plane of the body surface (Fig. 7d).

Fig. 8 shows the average result of 5 experiments. Curve (a) shows the additional sound pressure recorded by the microphone, when it has been surrounded by the body of a locust. Curve (b) shows the excess pressure on the diaphragm due to diffraction for the microphone, adaptor, and cathode follower alone (i.e. the free field response minus the pressure response under conditions without diffraction).

The total change in the sound pressure on the diaphragm due to diffraction on the locust plus microphone (curve c) is then equal to (a) plus (b). The curves (d) and (e) show the calculated effects for a cylinder and a sphere (Skudrzyk, 1954) of comparable size (radius 3.5 and 5 mm, respectively).

These experiments were easy to perform, but the results are difficult to interpret. The surplus pressure measured on the surface of the locust may be influenced by the presence of the adaptor and cathode follower. The influence should, however, be almost negligible at the surface opposite the adaptor (according to some measurements performed by Mr.

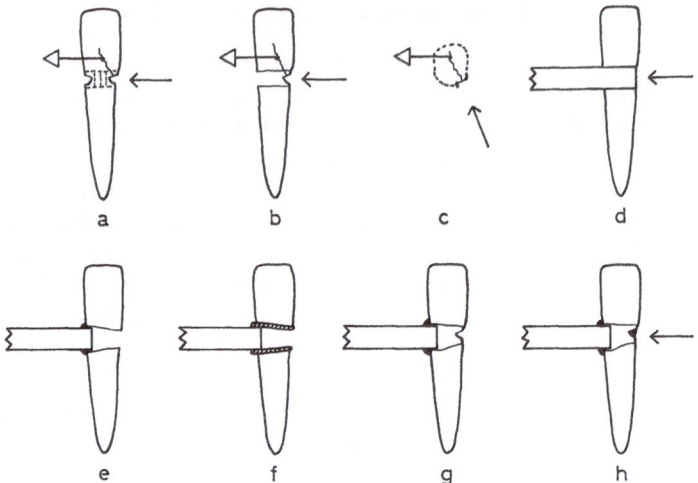

Fig. 7a–h. The different preparations used in the experiments. The figures a–c show the recording of nerve responses (a intact locust; b operated locust; c semi-isolated ear). The figures d–h show the measurement of relative sound pressures by means of a 6 mm microphone (d effect of diffraction; e tube resonances of operated locust (cf. b); f as e, but with plastic tube; g sound transmission with intact tympanum; h as g, but with the tympanum blocked with paraffin wax)

G. Rasmussen, Brüel & Kjär). Thus, curve (c) should be fairly close to the surplus sound pressure expected for an intact locust on the surface facing the sound source. At high frequencies the curves (d) and (e) approach 6 dB in an oscillatory manner. A similar behaviour may be expected for curve (c), but these oscillations are of no practical interest, since the ear is very insensitive above 40 kHz.

It should be emphasized, however, that from a theoretical point of view this experiment is unsound. A final solution to the problem will have to await experiments with a smaller microphone without adaptor, which can be buried in the body of the locust.

The curves (d) and (e) show the surplus pressure for a point on the surface facing the sound source. At the surface pointing away from the sound source very little change in sound pressure can be expected ("shadows" occur only at much higher frequencies).

2. Sound Conduction through the Locust Body

In the previous section the observed effect of diffraction was compared with the expected values for bodies with rigid surfaces. It is, however, not to be expected that the surface of insect bodies should be rigid when acted on by a sound wave. The sound wave will cause the

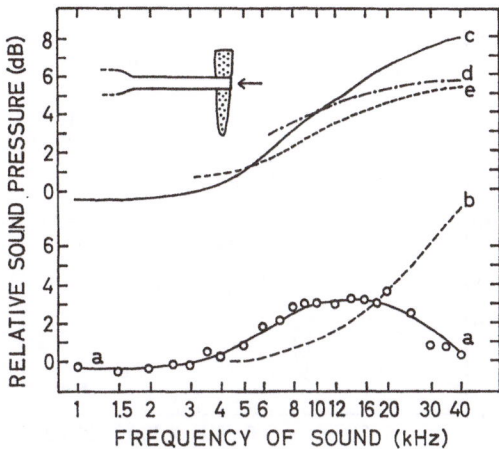

Fig. 8. The effect of diffraction on the sound pressure at the surface of a locust. Curve a: the additional sound pressure recorded by the microphone when surrounded by the body of a locust (average of 5 experiments). Curve b: the excess pressure due to diffraction for the microphone alone. Curve c: a plus b (= the total effect of diffraction for locust plus microphone). Curve d: the surplus pressure expected for an infinitely long, rigid cylinder of radius 3.5 mm. Curve e: the same for a sphere of radius 5 mm

cuticle to vibrate, and some of the sound will be transmitted through the insect. The sound transmission may be fairly efficient at some frequencies, but at other frequencies a large part of the sound may be absorbed by the cuticle or the internal tissues. Thus, one may expect both the surface material (cuticle) and the internal structures to act as acoustic filters. Some preliminary experiments were performed in order to estimate the magnitude of these effects.

In one series of experiments the locust was operated on as described above (Fig. 7b), and the 6 mm microphone was placed in the body behind the tympanum (Fig. 7g). The sound transmitted through the tympanum and cuticle was measured. The movements of the tympanal membrane were then hindered by means of wax (Fig. 7h) and the measurements were repeated. It is important to ensure a tight-fitting connection between the animal and the sides of the microphone; this was obtained with wax.

Fig. 9 shows the average result of four experiments. The points indicate the sound level recorded by the microphone, relative to the sound level on the surface of the animal. Thus, the values have been corrected for the surplus pressure at the surface due to diffraction (Fig. 8). Apparently, a fairly large amount of the sound energy is transmitted through

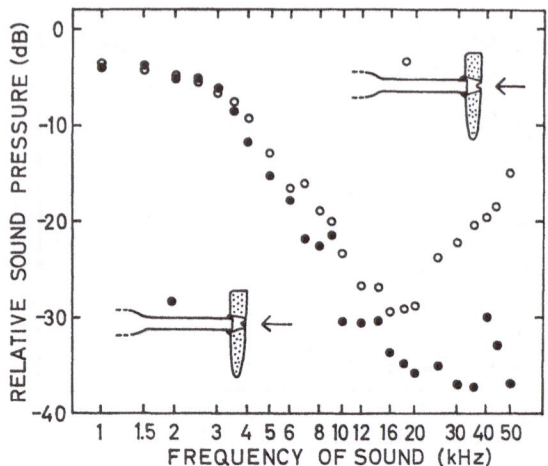

Fig. 9. The sound pressure inside the body of a locust relative to that outside (zero dB). ○ Intact tympanum (Fig. 7g). ● Tympanum blocked with paraffin wax (Fig. 7h). Each point represents the average of four experiments. The values have been corrected for the effect of diffraction (Fig. 8)

the cuticle at low frequencies (-4 dB is equal to 63%). At higher frequencies the sound transmission decreases rapidly (-20 dB $=10$%; -40 dB $=1$%). When the tympanum is intact (open circles), the transmission is a minimum around 16–20 kHz. The blocking of the tympanum (black dots) does not affect the sound transmission very much at low frequencies, but at higher frequencies the transmission is significantly less. This experiment explains why Autrum *et al.* (1960) did not observe any effect on the directivity at 7 kHz when blocking the opposite ear with wax.

These experiments showed the combined filtering effects of the cuticle and of the interior of the animal. In order to separate these two systems another series of experiments was performed, in which the ear and some of the surrounding cuticle had been removed (Fig. 7e). The experimental procedure was similar to that just described. In Fig. 10 the sound level relative to that outside (corrected for diffraction) is shown with black dots (average of 5 experiments). The sound level is now greater than outside the animal for frequencies up to about 10 kHz. It then drops off, and around 20 kHz it is about 12 dB below that outside.

This behaviour appears to be caused by a closed-tube resonance and a filtering effect of the internal tissues: After each of these experiments the microphone was placed at the end of a (hard) plastic tube of the same length and diameter as the hole in the animal. The tube was sur-

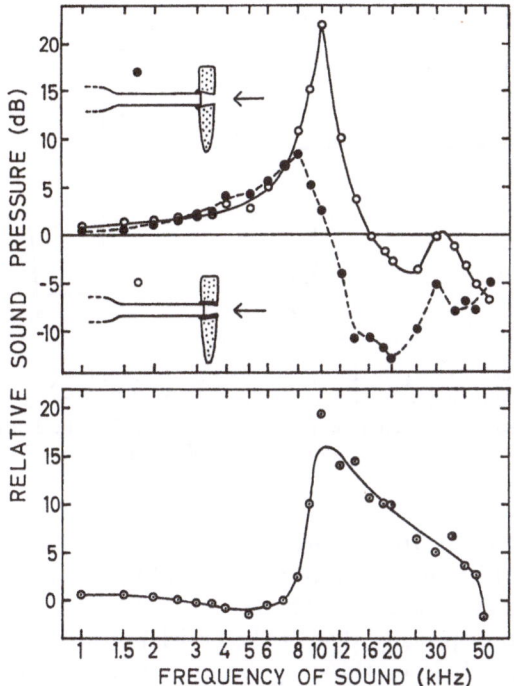

Fig. 10. Closed-tube resonances for an operated locust (●) and for a plastic tube surrounded by the body of a locust (○), cf. Fig. 7e and f. The lower curve gives ○ minus ●. The upper curves have been corrected for the effect of diffraction (Fig. 8) and are the average of 5 experiments

rounded by the body of the locust (Fig. 7f), in order to obtain the same effect of diffraction. The values thus obtained are indicated as open circles on Fig. 10. In this experiment the sound level was a maximum at two frequencies (about 10 and 30 kHz). At these frequencies the effective length of the tube (the real length plus a correction for the open end) is equal to one quarter and three quarters of the wavelength, respectively. Apparently, the observed maxima are due to the expected closed-tube resonances.

Both the curves on Fig. 10 have a maximum around 30 kHz. At low frequencies (up to 7–8 kHz) the curves are almost identical. It is therefore reasonable to conclude that the difference between the curves above 8 kHz reflects an absorption of sound in the internal tissues of the locust. It is apparent from the plastic-tube curve that some sound energy is lost at higher frequencies; the plastic-tube is, of course, not an ideal tube with rigid walls. Therefore, the difference between the curves (the

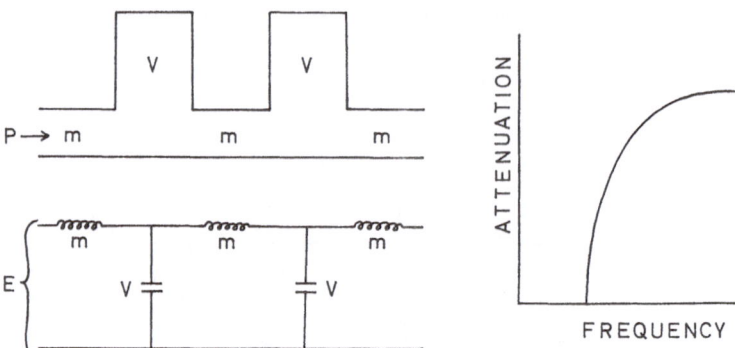

Fig. 11. An acoustic low-pass filter and its electrical analogy (left). At right the approximate attenuation curve is indicated. Modified from Beranek (1954). (V volume of air with springiness; m mass of air; P sound pressure, which is equivalent to the voltage E)

lower curve on Fig. 10) is only a rough estimate of the attenuation caused by the filtering action of the internal tissues of the locust. More careful studies are needed on this problem. Also, the dependence of the filtering action (especially that of the cuticle) on the angle of incidence should be studied, since this may be important for the directivity to low-frequency sounds.

Acoustic Filters. The physical basis for the filtering effect of the locust body is not known, but acoustic devices with approximately the same filtering properties are well known from technology. The behaviour of acoustic filters may be represented by means of an electrical analog. At low frequencies (the device considerably smaller than the wavelength of sound) the analogy to electrical filters may be fairly accurate. At higher frequencies, however, the devices will behave as transmission lines, and the analogy breaks down.

An acoustic low-pass filter (e.g. a silencer on a car) commonly consists of a tube, which has openings to closed chambers (Fig. 11). In the electrical analogy (Fig. 11) the mass of the air in the sections of the tube are represented by inductances, and the stiffness of the air inside the side chambers by capacitances. The electrical analogs for the sound pressure and velocity of the air particles are the voltage and current. The effect of friction has been neglected, but it can be built into the model as resistances.

The approximate attenuation curve for such a device (from Beranek, 1954) is shown in Fig. 11. As the frequency is increased, the attenuation increases abruptly. This is the same type of behaviour as that of the lower curve on Fig. 10. In the operated locust the remaining parts of the body contain a number of air spaces which—together with the tubular hole—might function as the device on Fig. 11. This idea is supported by the observation that the sound level changed a few dB in connection with each respiratory movement (this was especially the case in the experiments on the preparations illustrated on Fig. 7g and h). On the other hand, the walls of the internal filter in locust are not rigid, so the real situation is certainly far more complicated.

3. The Radiation Impedance

The radiation impedance (Z_r) of the front side of the membrane *in situ* is not identical with that calculated for the unbaffled membrane in paper II. An exact calculation is difficult, but the order of magnitude may be estimated by means of some simple approximations. At low frequencies (fundamental mode of vibration) the membrane can be regarded as equivalent to a spherical sound source of order zero. The surface of the sphere should be equal to the surface of both sides of the membrane, i.e. its radius should be $1/\sqrt{2}$ times the radius of the membrane (see Skudrzyk, 1954, p. 286). With this approximation the radiation impedance becomes

$$Z_r = R_r + j\,X_r = \pi\,a^4\varrho\,\omega^2/c + 4.4\,j\,\omega\,\varrho\,a^3 \qquad (11)$$

where $j = \sqrt{-1}$,

a = radius of membrane (m),
ϱ = density of air (about $1.2\,\text{kg/m}^3$),
ω = angular frequency (s^{-1}),
c = velocity of sound (ca. $344\,\text{m/s}$).

This equation is valid for $\omega\,a/c < 1$, i.e. up to about 50 kHz. Several other approximations can be used. For example the membrane may be regarded as equivalent to a plane, circular piston in an infinite baffle (see Beranek, 1954) or in the side of a rigid sphere (see Morse, 1948). The values thus obtained are all of the same order or magnitude.

In the present case the radiation mass becomes about 5 µg. This value should be compared with the weight of the thin part of the tympanum (about 9 µg). The resistive component (R_r) of the radiation impedance becomes about $4 \cdot 10^{-7}\,f^2\,\text{Ns/m}$ (f in kHz). This value should be compared with the value due to internal friction in the membrane (about $5 \cdot 10^{-5}\,\text{Ns/m}$ for the thin membrane). Apparently, R_r can be ignored at low frequencies (up to 10 kHz), but not at higher frequencies.

It should be remembered, however, that Eq. (11) is only valid for the fundamental mode of vibration. At higher modes the radiation will be proportional to the overall volume velocity of the membrane. The volume velocity decreases approximately as $1/n$, where n is the number of the mode of vibration (see Lax, 1944).

4. Expected Effect of the Backing on Membrane Vibrations

It is well known that the sound radiation from a loudspeaker backed by a closed box depends on the volume of the box and on the sound absorption properties of its interior. Similarly, the vibration of a membrane like the locust tympanum will depend on the acoustic properties of the tracheal air sacs and other tissues behind the ear.

The expected effects also depend on whether the travelling velocity of transverse waves in the membrane is smaller or greater than the velocity of sound waves in air. In the former case the movement of the individual parts of the membrane will quickly be transmitted to the entire volume of air behind the membrane and cause a homogeneous change of pressure. This change in pressure will affect other parts of the membrane without much delay. In the latter case the effects of a localized movement will be transmitted to other parts through the membrane and only much later through the air.

The velocity (v) of the transverse waves in the membrane can be found from

$$v = \sqrt{\frac{T}{\sigma}} \tag{12}$$

where $T =$ tension per unit length (N/m),
       $\sigma =$ mass per unit area (kg/m^2 = Ns^2/m^3).

In the present case v becomes 5–8 m/s, i.e. much less than the velocity of sound in air (344 m/s).

If the walls of the air space are rigid, the stiffness of the air may be added to the membrane tension. As a result, the resonance frequencies of the membrane increase. This increase may be fairly large for the fundamental vibration, but it can almost be ignored for the higher modes of vibration (because the higher modes have a much smaller total volume displacement and thus a much smaller effect on the pressure in the air space). The change of resonance frequency can be calculated fairly exactly, if the effective volume of air is known. In the present case, however, the walls of the air sacs are soft. Thus, it is doubtful whether calculations based upon measured volumes are realistic (see below).

The presence of "soft" material in the space behind the ear will contribute to the total friction in the system. One may therefore expect the damping of the membrane to increase, when fat is accumulating behind the ear. It is possible to calculate the magnitude of this damping, but here again the properties of the internal air spaces in a locust are so far from the ideal that calculations do not seem to be realistic.

5. Force and Phase in the Isolated and Operated Ear

The force acting to move the tympanum in the isolated ear and the phase lag of the resulting vibration can be calculated by means of Eqs. (8) and (9), if the effective distance (Δl) is known. It is, however, not possible to estimate Δl directly from the anatomy of the isolated ear. In the following we shall estimate these parameters by comparing the isolated ear with the operated ear described above. The operated ear is about 23 dB more sensitive at 3.5 kHz than the isolated ear. If the acoustical properties of the operated ear can be determined, we can use this value to calculate the force for the isolated ear. By doing so, we assume that the magnitude of the vibration at threshold is the same in both preparations.

In the operated ear in situ the effective distance $(\varDelta l)$ from the front side to the back of the membrane can be estimated fairly precisely from the anatomy as being about 11 mm. The force will be determined mainly by this distance, but the effects of diffraction, the transmission through and into the animal, and the radiation impedance of the membrane must also be considered.

The effect of diffraction is uncertain, but the experiments show (Fig. 8) that it can be neglected at 3.5 kHz. The attenuation in the duct leading to the back side of the ear is also negligible at 3.5 kHz. The closed-tube resonance in the duct will, however, tend to increase the sound pressure on the back of the membrane by about 2 dB (Fig. 10). The sound transmission through the cuticle (i.e. by other routes than the tympanal membrane) is fairly large at low frequencies (Fig. 9), but in the present calculation it has been neglected. The error thus introduced is not likely to be important, since the sound pressure on the back of the membrane will be dominated by the closed-tube resonance.

When the membrane of the operated ear is set into vibration, it will see the tubular hole in the body as a mixed mass-resistance element with the impedance (Beranek, 1954, p. 137)

$$Z = A^2 \left[\frac{\varrho\left(\frac{1}{a}+1\right)\sqrt{2\omega\mu}}{\pi\,a^2} + \frac{j\,\omega\,\varrho(1+0.6a)}{\pi\,a^2} \right] \tag{13}$$

where Z = mechanical impedance (Ns/m),
$\quad$ A = area of membrane (m²),
$\quad$ a = radius of tube (m),
$\quad$ π = 3.14 ... ,
$\quad$ ϱ = density of air (about 1.2 kg/m³),
$\quad$ ω = angular frequency $(= 2\pi\,f)$,
$\quad$ μ = kinematic coefficient of viscosity (about $1.56 \cdot 10^{-5}$ m²/s),
$\quad$ l = length of tube (m),
$\quad$ $j = \sqrt{-1}$.

The resistive term, which is somewhat dependent on frequency, is found to be about $2.5 \cdot 10^{-6}$ Ns/m at 3.5 kHz. This part of the impedance can be neglected, since the resistive part of the membrane impedance is about $5 \cdot 10^{-5}$ (for the thin membrane, see paper II). The mass calculated from the second term in Eq. (13) is about 4 µg, i.e. approximately equal to the radiation mass of the unbaffled membrane (paper II). This part of the impedance cannot be neglected.

Eq. (13) is valid when the effective length of the tube is much smaller than the wavelength of sound. At higher frequencies the behaviour is more complex. When the effective length of the tube is half the wavelength Z will be a minimum (cf. open organ pipes). Thus, one should expect a resonance at about 20 kHz, but because of the absorption in the body this resonance remains a theoretical possibility only. The radiation resistance of the front side of the membrane has been considered above. It can be neglected at 3.5 kHz.

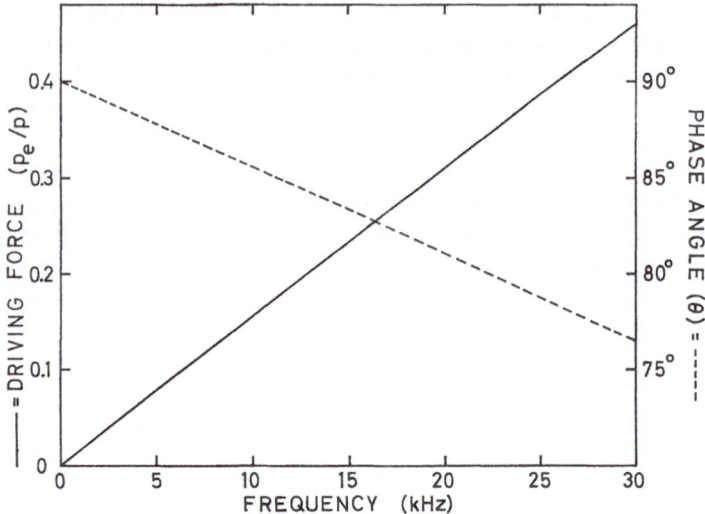

Fig. 12. The driving force (solid line) calculated for the isolated ear preparation indicated as a fraction of the force acting on a similar pressure receiver. Broken line indicates the phase angle between the driving force and the sound pressure

Thus, the force acting on the membrane in the operated ear will be determined by the 2 dB surplus pressure on the back of the membrane and an effective distance of 11 mm. By means of Eq. (6) one finds that the equivalent pressure (p_e) is 0.8 times p, i.e. the force is 0.8 times that in a similar pressure receptor.

This value can be used to calculate the force acting to move the membrane in the isolated ear. This preparation is on the average 23 dB (15 times) less sensitive than the operated ear. Therefore, p_e of the isolated ear is $(0.8/15)$ $p = 0.055$ p at 3.5 kHz. By means of Eq. (8) one finds $\alpha = 3.1°$ and $\Delta l = 0.85$ mm. This value can then be used to calculate the variation in force with frequency and the phase angle (Θ) between p and p_e. Both the force and the phase angle (Θ) appear to be almost linearly related to frequency in the frequency range of interest here (Fig. 12). These values were used in paper II to calculate the phase lags (in the experiments with the capacitance electrode) and the damping factor of the membrane. It is seen that $\Theta = 88.5°$ at 3.5 kHz, i.e. at low frequencies the phase angle between the sound pressure and the driving force is not far from the 90° expected for the pressure gradient itself [Eq. (3)].

Discussion

These experiments demonstrate that the sensitivity of the locust ear to low-frequency sound depends very much on the surrounding

Table. *The table summarizes the main characteristics (sensitivity, type of sound recei-*
ver, and the approximate magnitude of the driving force) of the three different ear
preparations studied. $p \cdot A$ *indicates the magnitude of the driving force in a pressure*
receiver (without diffraction). Further explanation in the text

Preparation	Threshold	Responding to	Driving force
Isolated ear			
around 3.5 kHz	35 dB	pressure gradient	0.055 $p \cdot A$
around 14 kHz	28 dB	pressure gradient	0.22 $p \cdot A$
Operated ear			
around 3.5 kHz	12 dB	(almost) pressure gradient	0.8 $p \cdot A$
around 14 kHz	27 dB	(almost) pressure	1.4 $p \cdot A$
Intact ear			
around 3.5 kHz	7–35 dB	?	1.2 $p \cdot A - p \cdot A$ (?)
around 14 kHz	28 dB	pressure	1.4 $p \cdot A$

structures. Three different preparations have been investigated. Some
of their main characteristics have been summarized in the Table.

The *isolated* ear is the simplest preparation, both from an anatomical
and from a physical point of view. It consists of the tympanum, the
groups of receptor cells (Müller's organ), and a supporting cuticular rim
(paper II, Fig. 6). Physically, it is a small, unbaffled membrane, and
it is likely to behave as a simple, symmetric pressure gradient receiver
over most of the frequency range of interest (1–40 kHz). The response
of single receptor cells in this preparation was reported in paper I. In
paper II it was shown that the frequency selectivity of the four groups
of receptor cells is based on the presence of two sets of resonances in
the tympanal membrane. In the present paper it is shown that the force
acting to move the tympanum at 3.5 kHz is about 0.055 times that in
a similar pressure receiver ($p \cdot A$). The force can be expected to increase
almost linearly with frequency (Fig. 12). Thus, the force acting to move
the membrane at 14 kHz is about 12 dB (4 times) larger than that at
3.5 kHz. The thresholds at 3.5 and 14 kHz are about 35 and 28 dB,
respectively.

The anatomy of the *operated* ear (Fig. 7b) is more complicated than
that of the isolated ear, but it is still more simple than the intact ear.
The operated ear is an asymmetric receiver; at low frequencies the
driving force is determined by the effective distance between the front
and back of the membrane and by the surplus pressure due to the closed-
tube resonance in the tubular channel behind the ear. With these assump-
tions the driving force at 3.5 kHz should be about 0.8 times $p \cdot A$. At
14 kHz the absorption of sound in the channel is so large that only a

small fraction of the sound reaches the back of the membrane (Fig. 10). The front side of the membrane, on the other hand, will experience a surplus pressure of about 3 dB due to diffraction (Fig. 8). The total driving force is therefore likely to be about 1.4 $p.A$. The threshold at 14 kHz is about 27 dB, i.e. almost the same as in the isolated ear. Thus, the sensitivity does not reflect the large difference in the magnitude of the driving force in these two preparations. One of the reasons for this probably is that the damping due to the resistive component of the radiation impedance at high frequencies is much larger in a baffled membrane (the operated ear) than it is in an unbaffled membrane (the isolated ear), see above and paper II. By chance the difference in damping may be just equal to the difference in driving force. It should be emphasized that the total damping at high frequences seems to be several times larger than that expected for Eq. (11) alone. This may indicate a considerable loss of sound energy to the internal tissues of the animal. Further studies are needed on this problem.

The resistive component of the radiation impedance is likely to affect the sensitivity of the ear near the resonance frequencies of the tympanum. The reactive component, however, will act as an addition to the effective mass of the membrane; this part is likely to cause a shift in the resonance frequencies (to lower values). The magnitude of this effect will be relatively moderate, since the resonance frequencies depend on the square root of the ratio between the total effective mass and tension [see Eq. (3) in paper II]. In the present case the larger radiation mass in the baffled membrane (to the front side and into the tubular channel) may be expected to cause a decrease in the fundamental resonance frequency of about 10–20%. It is interesting that two out of 9 operated ears were more sensitive to 3 kHz than to 3.5 kHz. More careful measurements are necessary, however, in order to determine the magnitude of this frequency shift.

1. The Acoustics of the Intact Ear

At high frequencies (above 10 kHz) the internal tissues of the locust act as an acoustic low-pass filter (Fig. 10). Additional low-pass properties were also found in the experiments in which the opposite ear was left intact (Fig. 9). Furthermore, the four air sacs between the ears may also act as a series of low-pass filters. Thus, in the intact ear very little high-frequency sound (10–20 kHz) reaches the back of the tympanum. It can therefore be concluded that for all practical purposes the intact (and operated) *ear acts as a pressure receptor to high-frequency sound* (i.e. to the frequencies capable of exciting the d-cells).

This concept has some interesting consequences. It means that a directivity to 10–20 kHz sound cannot be based on the direction of the vectorial component of sound waves; if present, it must be based on the (relatively small) differences in sound level due to diffraction. Also, this concept partly explains why the sensitivity of the d-cells is so little influenced by changes in the amount of fat between the ears (which in a pressure gradient receiver would be likely to affect the amount of sound reaching the back of the membrane and thus to affect the magnitude of the driving force).

The sensitivity of the ear to low frequencies (below 8 kHz) is markedly influenced by the amount of fat and other tissues between the ears (Figs. 4 and 5). The very meagre animal is the most sensitive of all preparations tested. The average sensitivity at 3.5 kHz is about 4 dB higher than that of the operated ear. This difference is probably due to the delay in sound transmission caused by the opposite tympanum and the air sacs: In the very meagre animal the interior can probably be regarded as composed of some resistances (the walls of the air sacs) connected to some capacitances (the stiffness of the air in the air sacs). The total arrangement forms a series of RC filters, which tend to delay the transmission of the sound wave through the animal.

The driving force at 3.5 kHz is about $0.8\,p \cdot A$ in the operated ear (see above). If the difference in sensitivity reflects a difference in the magnitude of the driving force, one finds (4 dB more than $0.8 =$) 1.2 times $p \cdot A$ for the intact ear, i.e. the driving force is somewhat larger than in a similar pressure receiver. This is, of course, possible in a pressure gradient receiver, if the difference in phase between the sound acting on the front and back of the membrane is larger than 60° (in the symmetric case).

It was mentioned above that the larger radiation mass in the (baffled) operated and intact ear causes somewhat lower resonance frequencies than in the (unbaffled) isolated ear. Similarly, in the intact ear the presence of a closed chamber behind the membrane will tend to increase the resonance frequencies, because the stiffness of the air will add to the stiffness of the membrane. In the operated ear the best frequency is probably between 3 and 3.5 kHz (see above). In contrast, four out of 13 intact ears were more sensitive at 4 kHz than at 3.5 kHz.

Thus, there is probably a small change in best frequency, but it is not likely to exceed 500 Hz. If the walls of the "box" behind the tympanum were rigid, a change of 500 Hz in the fundamental resonance frequency would correspond to a box volume of about 0.4 cm³ (see Morse, 1948, p. 194). The real volume of the air sacs between the ears is about 0.1 cm³, i.e. the "effective volume" of the air spaces is at least four times larger than the anatomical volume. This means that the walls of the air sacs must be very "soft" at the sound frequencies in question.

The stiffness of an air space is inversely proportional to its volume. In the very fat, intact locust the lumps of fat may cover most of the tracheal walls separating the air sacs. Thus, the "chamber" behind the membrane may anatomically be restricted to the air sac directly in contact with the tympanum. Consequently, one might expect a further increase of the fundamental resonance frequency when fat is deposited between the ears. No increase has been found in the present experiment (because of the limited number of observations, it is not possible to state that an increase is absent, but, if present, it must be very small). Here again, this observation demonstrates the soft nature of the walls of the interior air spaces. It might be interesting to place small chambers with known acoustic properties behind the ear and compare the results with those reported here.

The body wall and internal tissues offer very little resistance to very low frequencies (1–2 kHz), see Fig. 9. This means that the low-frequency sound reaching the back of the membrane is composed of a part travelling directly through the body wall (with a phase shift?) and a part coming through the opposite ear. It remains to be determined how this affects the hearing. In particular, the mechanism of directivity at very low frequencies may very well turn out to be fairly complicated.

It also remains to be determined whether—or to what extent—the intact ear is working as a pressure gradient receiver at frequencies around 3.5 kHz. In the very meagre animal there is no doubt that a considerable part of the sound reaches the back side of the membrane, but we cannot be sure how much sound is transmitted through the fat animal. It is apparent from the calculations performed above that the magnitude of the driving force at 3.5 kHz is fairly independent of the type of receiver: using the pressure gradient model the driving force becomes about 1.2 times $p \cdot A$, i.e. only about 20% different from the force in a pressure receiver. This result clearly demonstrates that the decrease in sensitivity in the fat animal has nothing to do with variations in the driving force. The low sensitivity must be due to the internal damping, which is probably caused by the softness and friction of the internal tissues.

The type of sound receiver is thus relatively unimportant for the study of driving forces. It is, however, of great importance for the biology of the animal, because it is linked up with the problem of directivity. At low frequencies the surplus pressure due to diffraction is extremely small (Fig. 8). Consequently, the locust should—in theory—be incapable of detecting the direction of the sound source, if its ears were acting as pressure receivers. (It can be assumed that the animal is incapable of using the difference in phase between sound reaching the two ears for detection of direction). The ability of the animal to locate a sound source is therefore likely to decrease together with the absolute sensitivity to low frequency sound, if the presence of fat between the ears causes a change in the type of sound reception.

A solution to these problems may be obtained by means of a careful study of the directional sensitivity of meagre and fat locusts at various frequencies. Such studies are at present being made by Dr. Lee Miller, Copenhagen.

2. Some Additional Factors

It was shown in part II (Fig. 7 on p. 72) that the compliance of the tympanal membrane is constant for deformations up to about 50–100 μm, i.e. in this range Hooke's law is obeyed. Therefore, small changes in pressure in the air sacs behind the tympanum are unlikely to affect the hearing. At deformations above 100 μm the force increases rapidly, i.e. the compliance decreases. In intact locusts the tympanal membranes move in connection with the respiratory movements. During heavy respiration the movement in and out may be some hundred μm. Therefore, one should expect a shift in the resonance frequencies of the tympanum (to higher frequencies) at the peak of each respiration. It is not possible to check this possibility by means of entire-nerve recordings, since bursts of spikes (from other mechanoreceptors ?) occur in the tympanal nerve in connection with the respiratory movements. It should, however, be possible to estimate the size of this effect (if present) by means of single cell recordings.

The cuticular rim supporting the tympanal membrane is very solid at most of the circumference of the membrane. At the ventral-anterior end of the membrane, however, the rim is fairly weak. It is possible to produce a slight change in the shape of the membrane by pushing and pulling the rim. Furthermore, in most species of grasshoppers one or two muscles attach to the rim. They have been named the tensor tympani muscles, but their function is not known. In Schistocerca, the tensor tympani attach to a stout cuticular protuberance of the rim, and although they are innervated, they cannot be caused to contract (Lee Miller, personal communication). These muscles may nevertheless have an effect on the ear in other species.

In general, the modes of vibration found in the isolated ear are also likely to occur in the intact ear. The resonance frequencies may, however, be slightly different, because of the different physical conditions in the two preparations. For most preferred frequencies a small shift will not mean much for the hearing ability of the animal. A few of the preferred frequencies are, however, caused by an interaction between two vibrations (one from each set). It was mentioned above (p. 118) that the addition of a box behind the membrane would be likely to affect the resonance frequency of the fundamental vibration more than those of the higher modes. This means that the patterns of interaction may not be exactly similar to those described in paper II. This question may be investigated by means of microelectrode recordings from single cells in the intact preparation.

I am most grateful to O. Juhl Pedersen, M. Sc. and Knud Rasmussen, M. Sc. for advice on the physical problems. My best thanks are also due to G. Rasmussen, M. Sc. for advice on diffraction problems, to Miss W. Taagerup for careful counting of about one million spikes, and to Professors Bent Christensen, F. Huber, and U. V. Lassen for discussions and pleasant working conditions. I am most indebted to the Deutsche Forschungsgemeinschaft and to Statens naturvidenskabelige Forskningsraad for generous support.

Appendix

Threshold and Threshold Criteria

During the present study an attempt was of course made to compare the results with those obtained by other investigators. It appeared to be a difficult task to judge the validity of the reported data, since there is a considerable scatter among the values reported by different authors using different techniques. Fig. 13 shows an example of this: The threshold curves for *Locusta migratoria* differ by about 45 dB. The effect of different amounts of fat between the ears (Fig. 4) may explain more than half of this difference, but about 20 dB must be due to differences in the experimental technique and in the definition of threshold. An attempt was therefore made to compare some of the commonly used methods. The comparison was performed with the assistance of some colleagues who were asked to judge whether a response to a stimulus was present or not.

The most commonly used method is the visual observation of an oscilloscopic trace. The threshold is defined as the stimulus intensity necessary to give a just noticeable response. This criterion is also used when the threshold is found by listening to an auditorially displayed signal. In both methods the detection of the signal required a certain amount of summation of the nerve impulses travelling in the auditory nerve. Therefore, these methods are biased in favour of the large groups of receptor cells: The thresholds determined at 14 kHz (where the signal is caused by about 10 d-cells) were about 5 dB "too high" in comparison with the threshold at 3.5 kHz (about 30 to 40 a-cells).

The auditory method is normally the more sensitive; using this method the threshold values determined were about 5–15 dB above those found with the method used in the present study. The thresholds obtained with the visual method depend rather much on the presentation of the signals: The amplification (size of the signal on the screen) and the sweep speed may each be responsible for differences of about 5 dB in the estimated values. Also, there was a considerable disagreement between the opinions of different observers. Unfortunately, most authors do not indicate exactly how they used the method.

In these two methods the response was observed at relatively low intensities, at which the nerve impulses had just started to form a summated response. The size of the large, summated "on-response" at higher intensities may also be used to estimate the relative sensitivity, e.g. in studies of directivity. The relationship between the sound intensity and the size of the "on-response" is sigmoid; it is therefore not possible to define a threshold value as the intersection of such a curve with the abscissa.

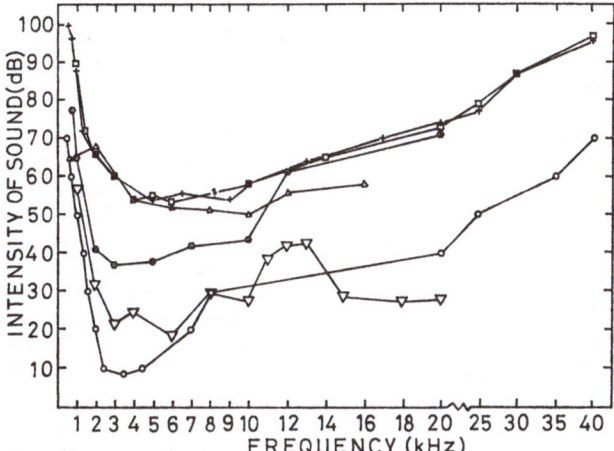

Fig. 13. Threshold curves for the tympanal nerve of *Locusta migratoria* determined by various authors

Two different threshold criteria have been used in my investigations. In the recordings from single receptor cells (paper I) the threshold was defined as the intensity necessary to give an average response of one spike more than the spontaneous activity to a sound pulse of 100 msec duration. Recordings were made from the entire nerve of isolated ears, and the threshold were compared with those determined from the single receptor cells (paper I, Fig. 11). The difference was about 11 dB at all frequencies. Thus, these methods are directly comparable, and they are not biased in favour of the large groups of receptor cells.

It should be emphasized that the threshold values thus defined are quite arbitrary. They are, of course, not more "true" than the values found by means of other methods, but the precision of determination makes them especially suited for studies of the biophysics of hearing. The relationship between the defined threshold in peripheral nerves and the "central threshold" determining behaviour has been discussed in detail by Roeder (1966).

Radar Studies. The problems involved in the detection of threshold signals by a human observer have been studied during the development of radar. The detection of signals on radar screens has many features in common with the detection of the threshold for the summated response to a stimulus in electrophysiological recordings. In both cases a human observer has to decide whether a response is present or not.

In the radar studies (see Lawson and Uhlenbeck, 1950) comparisons were made between an aural method (listening to the signal), visual methods (observing the signals on a normal sweep oscilloscope or on an intensity modulated oscilloscope), and various meter deflection systems. It was found that these methods are equivalent only if each method is pushed to the limit of its capacity. In practice this will almost never be done, and therefore the different methods will usually give different signal thresholds.

A result of great interest to biologists was that consistent results could not be obtained when the observers were asked to estimate the minimum detectable signal. The individual estimates differed by so large factors that it was nearly impossible to obtain reliable averages. Also, the estimates of a given observer changed considerably from day to day and depended on the setting of the apparatus (e.g. trace intensity, focus conditions, and contrast in visual observation), on previous training, and on the observer's state of mind.

Thus, psychophysiological properties of the human senses and psychological factors such as attention and motivation tend to make direct observation by humans inaccurate and biased. In theory, the essential limitation for the detectability of a signal is due to the statistical nature of the problem, but this is only true when the function of the observer is reduced to measuring or counting.

The method used in the present experiments can only be employed when the spontaneous activity in the nerve is so small that summation is rare. Unfortunately, this is not often the case in nerve recordings. In larger nerves the objective method developed in the radar studies may be found useful: In this method the position of the signal is varied at random, and the signal is made to appear at one of six positions on the screen. The observer is not asked to judge whether or not the signal is present, but only to guess its position. In this way the judgment is simplified, and psychological factors are eliminated. There is still a considerable statistical scatter, but the magnitude of the response can be estimated with a surprising accuracy (e.g. with about 100 observations the relative signal power can be estimated within 0.5 to 1 dB).

Although this method is far more accurate than those normally used by biologists, it is still biased in favour of the large groups of receptor cells and/or in favour of the larger axons in a compound nerve. Even when the biologist has access to an averaging computer, this limitation is unescapable.

References

Autrum, H.: Über Lautäußerungen und Schallwahrnehmung bei Arthropoden. II. Das Richtungshören von *Locusta* und Versuch einer Hörtheorie für Tympanalorgane vom Locustidentyp. Z. vergl. Physiol. **28**, 326–352 (1941).
— Schwartzkopff, J., Swoboda, H.: Der Einfluß der Schallrichtung auf die Tympanal-Potentiale von *Locusta migratoria* L. Biol. Zbl. **80**, 385–402 (1961).
Beranek, Leo L.: Acoustics. New York: Mc-Graw-Hill 1954.
Lawson, J. L., Uhlenbeck, G. E.: Threshold signals. New York: McGraw-Hill 1950.
Lax, M.: The effect of radiation on the vibrations of a circular diaphragm. J. acoust. Soc. Amer. **16**, 5–13 (1944).
Michelsen, A.: The physiology of the locust ear. I. Frequency sensitivity of single cells in the isolated ear. Z. vergl. Physiol. **71**, 49–62 (1971).
— The physiology of the locust ear. II. Frequency discrimination based upon resonances in the tympanum. Z. vergl. Physiol. **71**, 63–101 (1971).
Morse, Ph. M.: Vibration and sound. New York: McGraw-Hill 1948.
Pumphrey, R. J.: Hearing in insects. Biol. Rev. **15**, 107–132 (1940).
Roeder, K. D.: Acoustic sensitivity of the noctuid tympanic organ and its range for the cries of bats. J. Insect. Physiol. **12**, 843–859 (1966).
Skudrzyk, E.: Die Grundlagen der Akustik. Wien: Springer 1954.

Axel Michelsen
Zoological Laboratory
Universitetsparken 15
DK-2100 Copenhagen 0, Denmark